Designing Sound

TMI TECHNIQUES of the MOVING IMAGE

Volumes in the Techniques of the Moving Image series explore the relationship between what we see onscreen and the technical achievements undertaken in filmmaking to make this possible. Books explore some defined aspect of cinema—work from a particular era, work in a particular genre, work by a particular filmmaker or team, work from a particular studio, or work on a particular theme—in light of some technique and/or technical achievement, such as cinematography, direction, acting, lighting, costuming, set design, legal arrangements, agenting, scripting, sound design and recording, and sound or picture editing. Historical and social background contextualize the subject of each volume.

Murray Pomerance
Series Editor

Jay Beck, *Designing Sound: Audiovisual Aesthetics in 1970s American Cinema*
Wheeler Winston Dixon, *Death of the Moguls: The End of Classical Hollywood*
R. Barton Palmer, *Shot on Location: Postwar American Cinema and the Exploration of Real Place*
Murray Pomerance, *The Eyes Have It: Cinema and the Reality Effect*
Colin Williamson, *Hidden in Plain Sight: An Archaeology of Magic and the Cinema*
Joshua Yumibe, *Moving Color: Early Film, Mass Culture, Modernism*

Designing Sound

Audiovisual Aesthetics in 1970s American Cinema

JAY BECK

RUTGERS UNIVERSITY PRESS

NEW BRUNSWICK, NEW JERSEY, AND LONDON

LIBRARY OF CONGRESS CATALOGING-IN-PUBLICATION DATA

Beck, Jay 1966–
Designing sound : audiovisual aesthetics in 1970s American cinema / Jay Beck.
pages cm
Includes bibliographical references and index.
ISBN 978–0–8135–6414–2 (hardcover : alk. paper) — ISBN 978–0–8135–6413–5 (pbk. : alk. paper) — ISBN 978–0–8135–6415–9 (e-book (web pdf)) — ISBN 978–0–8135–7307–6 (e-book (epub))
1. Sound motion pictures. 2. Motion pictures—Production and direction—United States—History—20th century. I. Title.
PN1995.7.B43 2016
791.4302'4—dc23

2015028626

A British Cataloging-in-Publication record for this book is available from the British Library.

Visit our website: http://rutgerspress.rutgers.edu

Manufactured in the United States of America

CONTENTS

ACKNOWLEDGMENTS

This book would have been impossible to complete without the assistance of Val Almendarez, Vanessa Theme Ament, Tim Anderson, John Belton, Dana Benelli, Marisa Carroll, Norma Coates, Don Crafton, Adrian Danks, Carol Donelan, Ofer Eliaz, Clark Farmer, Gregg Flaxman, Ian Gardiner, Tony Grajeda, Dennis Hanlon, Helen Hanson, Michele Hilmes, Laska Jimsen, Kathryn Kalinak, Anahid Kassabian, Chris Keathley, Mark Kerins, Jim Lastra, Franck Le Gac, Allison McCracken, Jason Middleton, Gregg Millman, Christopher Polt, Angelo Restivo, Ron Rodman, Vicente Rodríguez Ortega, John Schott, Jeff Smith, Deborah Tudor, Cathy Weingeist, Elisabeth Weis, Joe Wlodarz, and Steve Wurtzler.

Thank you to Ioan Allen/Dolby Labs, Mark Berger, Les Fresholtz, Gary Katz, Tom Kenny/Mix Magazine, Paul Rayton/American Cinematheque, Dave Stone, Frank Warner, and Jim Webb for granting me access to their work and their thoughts about film sound. Special thanks to Larry Blake, Paul Soucek, and Randy Thom for opening many doors in the world of professional film sound. And a very special thank you to Walter Murch and Richard Portman for giving me so many stimulating ideas.

A large portion of this book was researched at the Margaret Herrick Library of the Academy of Motion Picture Arts and Sciences, the Louis B. Mayer Library of the American Film Institute, the Theatre Division of the New York Public Library, and the Celeste Bartos Film Library at the Museum of Modern Art. Documents exclusive to a particular collection have their source listed in notes. Funding for this research was provided by grants and awards from the University of Iowa Student Government, the Robert Olney Travel Award, the Seashore Fellowship of the Graduate College of the University of Iowa, the DePaul University Research Council Competitive Research Grant, and sabbatical research leave from Carleton College.

A portion of chapter 5 appeared in an earlier version as "Citing the Sound: *The Conversation, Blow Out,* and the Mythological Ontology of the Soundtrack in '70s Film" in the *Journal of Popular Film and Television* 29, no. 4 (Winter 2002).

Thank you to Leslie Mitchner, Murray Pomerance, Carrie Hudak, Lisa Boyajian, Katie Keeran, Eric Schramm, and everyone at Rutgers University Press. Finally, I would like to thank Corey Creekmur, Kembrew McLeod, Leighton Pierce, and Lauren Rabinovitz for their guidance, and my deepest thanks to Rick Altman and Cecilia Cornejo for their insight and inspiration.

Designing Sound

1

Introduction

The State of the Art

Film sound in the late 1960s and early 1970s was a major component of an organized attempt on the part of several filmmakers to challenge the conventions of Hollywood cinematic practices through formal and narrative experimentation. With the weakening of the major studios in the wake of several economic and social factors (the 1948 Paramount decree, the fallout from the HUAC hearings, the rise of television and subsequent drop in cinema attendance, the resistance in foreign markets due to quotas on American films, the challenges to and demise of the Production Code), the Hollywood studio system underwent a series of radical changes from the late 1950s through the 1970s. Commonly seen as a period of economic instability (the closing of RKO studio and bankruptcy of Twentieth Century–Fox) and commercial failures (*Cleopatra, Darling Lili, Star!, Tora!, Tora!, Tora!*), the late 1960s and early 1970s are often elided in cinematic histories that prefer to emphasize a link with the prior system in the rise of the blockbuster and the sequel strategies of the late 1970s. Even theorists who examine the period often do so by treating it as an anomalous moment, a temporary break in the relatively stable timeline of Hollywood cinematic representation. One of the decade's staunchest defenders, Robert Kolker, bracketed these cinematic changes by noting, "That brief moment of freedom . . . was really a freedom to be alone within a structure that momentarily entertained some experimentation."[1]

However, a closer study of the period reveals a turmoil that was less random than previously perceived. During the late 1960s and early 1970s there was a break in the homogenous output of the film industry as studio styles, genres, and narrative centrality weakened along with the industry. This gave filmmakers the freedom to explore new forms of representation in their chosen images and sounds, and it is these experiments that form the focus of this study. While the visual aspects of cinema during the 1960s and 1970s are relatively well documented, there is a dearth of material evaluating the impact of the changes in film sound. Journals such as *American Cinematographer* paid near-exclusive attention to the visual components of Hollywood cinema during this time (despite the fact that almost half of their advertisements related to audio technologies) with only the slightest mention of film sound's significance. In fact, sound was routinely ignored, and even marginalized, in most writings about cinema during the 1960s and 1970s until the introduction of new "spectacular" technologies, including Sensurround and Dolby Stereo.

Cinematic practices in the 1967–1979 period, and specifically film sound practices, served as a site for negotiation of an American identity through the reconfiguration of outdated models and forms of representation and their alignment with more radical political and social perspectives. Industrial, technological, and aesthetic factors must be seen in light of the ideological constructs of the period for a full understanding of the changes that occurred in 1970s film sound. The primary value in examining this period is an understanding of the potential pathways that cinema could have followed, and still can follow, in its sound practices. The changes ushered in by the near universal acceptance of Dolby Stereo before the end of the 1970s essentially shut down the experiments of the first half of the decade and obscured the aesthetics pioneered by those films. Following the model of "crisis historiography," as delineated by Rick Altman in his book *Silent Film Sound*, this study examines the period's identity crisis in filmmaking, which resulted from changes in industrial standards.[2] It reveals that deviations were partially instigated by pressures from within the film industry to adopt and assimilate aspects from parallel industries. These new models of cinematic presentation offer an important site for inquiry into changes in the cinema system.

As one form of experimentation for late 1960s filmmakers, film sound offered an untheorized and relatively unchanged set of practices that were inherited artifacts of the studio system of production. The fact that filmmakers chose to manipulate, abstract, and reconfigure practices of sound recording and mixing during this period shows not only a willingness to break free from the restrictions of the studio system but also a drive to change audience perception. During the late 1960s sound aestheticians explored new methods of constructing film sound tracks in an attempt to rethink regimes of seeing and hearing in narrative cinema. Formal alterations appeared in multitrack mixing, new miking strategies, the use of location sound in lieu of looped dialogue, a reintroduction of stereo, and the dismantling of hierarchically structured systems of film sound editing and mixing. Filmmakers resisted models that dictated accepted industrial norms of how to make a film correctly and proceeded to challenge audiences with films that required spectator/auditors to engage the cinematic action on new, visceral levels. Models were drawn from television, the music and recording industries, radio, and even surveillance equipment in an attempt to develop new forms of representation that addressed disparate and specialized audiences. Crucially, the narratives of many of these films relied on the cultural alienation and uncertainty felt by both the filmmakers and their target audiences in the late 1960s. These changes made the gap between the uniform address of classical Hollywood and the emergent cinematic forms of the period more palpable.

Most studies of film in the 1970s tend to look at the decade as a departure from the confusion wrought by the decline of the studio system into the organization and structure of blockbusters, sequels, and multi-generic crossover films. In this kind of history, the films of the 1967–1979 period offer very little by which to gauge the progress of the romantically mythologized Hollywood cinema. There is a tendency for historians to overlook this period, bookending it with cursory mentions of *Bonnie and Clyde* as the death knell of the studio system, and the reemergence of a "New Hollywood" with *The Godfather*, *Jaws*, and *Star Wars*. With the exception of scholars like Robert Kolker and Robin Wood, there is a paucity of critical work that treats the films from this period as crucial to the development of cinema.[3] The vast majority of film histories position these films as predecessors to the commercial successes of the following decades without exploring the formal and aesthetic changes that they introduced. What is needed is a methodology for examining this period that can shed new light on the practices that emerged and were assimilated by cinema as an industry, and those that stood as idiosyncratic experiments in altering cinematic form. Only through a study of these moments can the full potential of the period be unearthed.

Sound functions as the central structuring device in this project, yet this book is not *exclusively* about sound—sound is also used as a heuristic to tell us more about cinema as an event. It is mobilized to elucidate several of the salient issues about cinematic structure in the period under scrutiny. Sound functions to allow access to three separate levels of understanding of the cinema system. First, it serves as an access point to explore and understand how cinema functioned during a certain era. Second, it provides an object of study that allows for the systematic analysis of cinematic spectatorship. And third, it offers a tool for reconceptualizing and rethinking cinematic practices today. Ultimately, this study of film sound from 1967 to 1979 is designed to tell us more about how we perceive cinema in the present. The study of film sound is a relatively new academic discipline compared to the extensive analysis of the image component of cinema; however, this absence of academic sensitivity to sound makes it no less important. By exploring what Altman calls "sound's dark corners" we are able to learn more about the function and place of cinema in contemporary society.[4] Here I contend that an examination of film sound reveals the potential of cinematic practice and, more widely, spectatorial roles.

In this period of sound experimentation, roughly from 1967 to 1979, the film industry underwent a massive change in industrial, technical, and aesthetic practices that were in keeping with a growing sense of discomfort in the public sphere. After the scrapping of the Production Code and its replacement with the rating system in 1968, cinema was free to explore modes of expression that challenged the fixed representational patterns of classical Hollywood. Sex and violence became visual markers of these changes, but the acoustical markers are harder to identify. While image strategies during the 1960s included greater use of direct lighting, hand-held cameras, zoom lenses, and the acceptance of lens flare as an aesthetic trope (devices that were considered antithetical to cinematic realism in prior decades), parallel changes in film sound lagged far behind. It was considered aesthetically acceptable to have grainy stock, oddly framed images, and a desaturated image, but insistence on narrative intelligibility of scripted lines still meant that almost all dialogue was heard clearly and recorded at close range. In the 1960s, with the introduction of technologies including lightweight Nagra magnetic tape recorders, smaller lavalier microphones with radio transmitters, graphic equalizers, and multitrack mixing boards, film sound could be reconceptualized and remobilized. This did not happen all at once, of course, and many of the changes that did occur were structured around the existing regimes of audiovisuality (lavalier mics and radio packs were used to get "clean" dialogue recordings at a distance), but the evolution in technology allowed several filmmakers to mobilize film sound toward creative ends.

The most famous example is Robert Altman's use of overlapping dialogue in his films of the 1970s. While overlapping dialogue was not new to Hollywood filmmaking—as heard in the screwball comedies of the 1930s—Altman revitalized the technique by multiplying the number of speaking voices and divorcing them from their spatial relation to the frame. As a deliberate byproduct of the lavalier microphone and radio pack, voices in Altman films are heard "directly," without reverberation, regardless of their proximity to the camera or their location within the frame. This odd equality of voices meant that spectator/auditors were able to follow multiple simultaneous conversations in a single scene. This freedom of interpretation opened up Altman's films to a new level of narrative complexity that was hitherto unknown in Hollywood filmmaking.[5] To ensure that the plurality of voices was heard in the release version of his films, Altman accommodated a plurality of voices in the production process through both multitrack production recordings and the democratization of the recording team. He conceptualized his production team as a collaborative endeavor rather than as a hierarchical construction passed on from the standardized patterns of Hollywood sound recording and mixing. This broke with classical film sound production standards to ensure a consistency of Altman's artistic vision while allowing the director to oversee the creation of the sound track in its entirety.[6]

Altman proved to be a significant creative voice in his mainstream studio films and smaller independent productions, but many directors working within the Hollywood system—such as Francis Ford Coppola, Martin Scorsese, Arthur Penn, and Terrence Malick—were also able to utilize the film sound track as an effective medium for expanding the dimensions of the narrative. Coppola foregrounded the plasticity and constructed nature of the sound track when he demystified the mixing process in *The Conversation* (1974). He, along with sound re-recordist Walter Murch, deflected audience expectations of easily comprehensible dialogue with the mumbled, whispered lines in both *Godfather* films (1972, 1974). Similarly, Scorsese worked with live, improvised dialogue and refused to re-record lines to provide better audience comprehension. As an ancillary result, there is an increased sense of offscreen space in Scorsese's films due to the realistic reverberant sounds that permeate his scenes. Scorsese also mobilized popular music as a register for internal character subjectivity, providing extradiegetic commentary on the profilmic events. Penn's sound tracks emphasized the physical aspect of violence in his films by mixing gunshots, explosions, punches, and other corporeal sounds at a much higher level than the surrounding dialogue. The result is an immediacy in the sounds that gives them the same visceral impact as their correspondent onscreen actions. On the opposite end of the scale is Malick, who revels in the ambient sounds of his films and their ability to allow the diegetic space to spill into the theater. It is not surprising that Malick readily adapted Dolby Stereo's ability to surround the audience in ambient sound while keeping the dialogue and internal narrations restricted to the screen speakers. These directors, as well as several others working in Hollywood, created a variety of competing models of sound usage that strained the boundaries of traditional recording and mixing practices, providing a tool for them to express their doubts and misgivings about American society. Technologies from other industries were modified and adapted to the exigencies of these experiments, and the period saw the growth and deployment of a variety of contrasting models of film sound usage.

But in the late 1970s, concurrent with films like *Star Wars* (1977), Dolby Stereo reintroduced classical rules of film sound recording and mixing that effectively served to patch over the gap created by prior sound experiments. *Star Wars* engendered an idealized sense of prosperity and validation of steadfast heroism—a striking visual and narrative contrast to the pessimistic films and "unmotivated heroes" that preceded it[7]—and it came wrapped in an acoustic package that introduced Dolby Stereo to a new generation of film viewers and makers. Dolby Stereo, the product of Dolby Laboratories, was a new procedure for mixing and encoding multichannel sound in the same space as the monophonic sound track on 35mm film. It marked a moment of promised potential for sound mixing practices, but at the same time it represented a specific, concerted effort to ensure that such potential never emerged, by virtue of its need for standardization in the recording and mixing processes. The system's backward compatibility meant that mixing practices had to be standardized and required that dialogue was always mixed in the center channel to ensure comprehension. By separating out dialogue mixing and elevating it to the top of the post-production sound hierarchy, forcing sound effects and music to be mixed according to prescribed methods, Dolby Stereo was a retreat from the democratic configuration of the sound team in the early 1970s to a classically minded hierarchy of mixing practices. It traded innovation for the commercial viability of audiovisual spectacle, linear narrative, and dialogue-driven stories. The acceptance of Dolby Stereo as a standard by the end of the 1970s meant that Hollywood cinema was unable to break completely with the classical model in order to build on the promise of the sonic experiments of the early 1970s.

The changes that took hold in the latter half of the 1970s, specifically the introduction of Dolby Stereo and its regimented mixing strategies, introduced a "new classicism" into Hollywood and derailed many of the formal and aesthetic changes initiated in the preceding years. Film sound shifted again from a medium capable of carrying an ideological message to a simple series of acoustic attractions. The literal "whiz-bang" function of Dolby Stereo allowed the sound track to expand into the three-dimensional space of the theater, but only after ensuring the fixity of the voice within the plane of the motion picture screen. These changes in film practice reveal an awareness of a higher order, an understanding that the standards and rules of filmmaking were just as constructed as the films themselves. In rethinking and revising these rules, filmmakers of the period sought to revitalize cinema as a site for social commentary and change. The ideological valences of the films reflected a sense of discontent with both the social order at large and its manifestation in the highly structured studio system of labor. During this brief period of creativity, filmmakers were temporarily able to free the sound-image relationship from the biases of technological and narrative determinism. A study of the films of this period is crucial to demonstrating the aesthetic potential of multichannel sound before the conservative strategies of spectator positioning in Dolby Stereo reinstated the classic divisions of labor and concretized the rules of narrative dominance over the interplay between sound and image. Therefore, this book foregrounds the specific historical factors that shaped the Hollywood system during this time, and the elements within the system that resisted the established forms of representation.

Any examination of the aesthetics of film sound in the 1970s requires understanding the dialectical nature of technological and industrial changes. Because aesthetics developed in accord with, as well as in reaction to, technological and industrial shifts, each aesthetic choice must be evaluated against the larger backdrop of cinematic evolution. For example, labor organizations, such as IATSE Sound Local 695 and Editors' Local 776, were relatively slow to respond to new roles in sound creation and regularly resisted technological advances under the guise of protecting the working rights of their constituents. Despite these strong barriers to change within the industry, several significant advances did occur in film aesthetics that were to shape the future direction of film sound in particular and film form in general. More than just singularities on the part of independent directors, these experiments of sound usage were part of a wide-ranging series of formal explorations undertaken by various directors who had become disenchanted with the rigid form of American cinema. In the collaborative production processes fostered by Arthur Penn and Robert Altman, the turn away from dialogue in the films of Monte Hellman and Terrence Malick, and the emergence of new sound roles in the films of Francis Ford Coppola and George Lucas, the 1970s represents a renaissance of cinematic aesthetics.

Viewed on the surface, the aesthetic changes that took place during the late 1960s and 1970s were a brief period of heterogeneity within an otherwise tightly regulated and controlled system. According to Douglas Gomery, in an article examining the state of the American film industry from 1983, the power structure of the major Hollywood studios was still firmly in place and there had been virtually no change in the large-scale economics of the industry. While recognizing the influence of many new directors and the rise of auteurism, he concluded that "little changed in the American film industry during the Seventies, despite all the pundits' claims of a 'new Hollywood.'"[8] More recently, Kristin Thompson updated a number of the assertions made in her coauthored tome on film style in American film, *The Classical Hollywood Cinema*. In *Storytelling in the New Hollywood*, Thompson set forth to prove empirically that the models of classical narrative structure outlined in *The Classical Hollywood Cinema* were still at work at the end of the twentieth century. Her argument centers on countering the claims made by Justin Wyatt, Thomas Schatz, Richard Maltby, and others that in the late 1970s the American cinema entered a "post-classical" era of fragmented, incoherent, and action-laden texts. What is curious about her rebuttal of these theories is the way in which it conveniently sidesteps the narrative developments and character deployments of the films of the early 1970s. She brackets off the decade succinctly: "I would suggest that the youthquake/auteurist films of the period from 1969 to 1977 or so were not harbingers of a profound shift in Hollywood storytelling but a *brief detour* that has had a lingering impact on industry practice."[9] Other than examinations of *Jaws* (1975) and *Alien* (1979), Thompson picks her examples from the 1980s and 1990s exclusively to demonstrate a near complete coherence between existing rules of classical Hollywood narrative construction and the form of storytelling used in contemporary cinema. However, she neglects to reconsider the films before this period, roughly from 1967 to 1975, which simply cannot be reduced to the constraints of the classical Hollywood narrative. Specifically, films like *Nashville, Night Moves, Two-Lane Blacktop, The Conversation, The Parallax View, Point Blank, Petulia, Klute, Mean Streets*, and *McCabe & Mrs. Miller* are conspicuously absent from her examination. Each of these films deploys its narrative around unmotivated heroes, antiteleological structures, and, occasionally, an ensemble cast without established stars. In numerous ways, there were very important changes occurring on the level of storytelling that Thompson was either unwilling or unable to integrate into her overarching theory. Most importantly, these changes in narrative form manifested themselves through new narrative modes that reconfigured the function of sound and image in American cinema. It is these films that demonstrate how film sound became a main tool for directors to redefine the rules of cinematic construction and to develop a new vocabulary for narrative cinema.

A deeper survey of this decade proves that not only were cinematic conventions challenged and redeployed for newfound uses, but many of these experiments still hold promise for the future of filmmaking. The following chapters examine a number of unique films and filmmakers who actively sought to change the landscape of American film during the late 1960s and 1970s. In each chapter, the choice of films and filmmakers reflects a unique approach to the question of film sound use and the possibility for either adapting previous techniques to new uses or the development of entirely new procedures. Each case study represents different aspects of film sound aesthetics in the 1970s, both successfully integrated aesthetic approaches as well as a number of less successful but important techniques. Significantly, nearly every example combines an increased awareness of the role of sound technicians and the enhanced recording and mixing technology that allowed sound practitioners to advance the syntax of film sound in their pictures. This is true despite wide variations in the working methods of some of the filmmakers. Certain directors like Francis Ford Coppola and Robert Altman preferred to operate by granting autonomy to the sound team and offering suggestions only during the final mix. In contrast, others like Alan J. Pakula and Martin Scorsese would take strong roles in post-production, directly supervising all aspects of a film's sound. Irrespective of their chosen method, these directors and their collaborators reconstructed the sound of American cinema in the late 1960s and the 1970s. What follows is a compendium of advances in the use of sound in American motion pictures, commensurate changes in sound practices, and the resultant aesthetics that emerged during the decade.

PART ONE

General Trends (1965–1971)

2

The British Invasion

In the 1960s, a confluence of new stylistic strategies gradually made its way into Hollywood filmmaking practices. With the auteur theory reviving interest in several American sound mavericks (Orson Welles, Alfred Hitchcock, Rouben Mamoulian, Ernst Lubitsch, etc.), new methods of sound experimentation, imported via British directors working in Hollywood, filtered their way into the mainstream. Despite the innovations brought into cinema by the first generation of sound pioneers and the brief "frozen revolution" of the 1950s magnetic audio sound,[1] much of sound practice in the 1960s was still rooted in a strongly hierarchical labor system and controlled set of production rules. Instead of the creative innovations of 1930s, Paul Monaco argued that in the 1960s, "Hollywood sound work during the decade remained a craft widely considered subsidiary to the main elements of production."[2] Although technically correct, it was in the 1960s that the seeds of creative sound use were sown, which came to fruition in the 1970s.

American cinema in the 1960s was in crisis as box office receipts were in steady decline and the technological innovations of the 1950s—widescreen, color, and stereophonic sound—failed to regain the audience lost to television. The first strategy employed by the studios was to follow a "business as usual" approach by relying on established genres, stars, and directors to entice viewers back to theaters. While this accounted for some major successes for the musical and epic genres like *The Sound of Music* (Robert Wise, 1965) and *Doctor Zhivago* (David Lean, 1965), it ultimately led to numerous flops such as *Doctor Dolittle* (Richard Fleischer, 1967), *Star!* (Robert Wise, 1968), and *Tora! Tora! Tora!* (Richard Fleischer with Toshio Masuda and Kinji Fukasaku, 1970) that nearly bankrupted the industry.[3] In response to the hemorrhage of money from underperforming pictures, Hollywood developed a second strategy by looking outside the film industry for an infusion of new blood and the biggest sea change since the 1930s. New talent was recruited from a variety of sources, and with them came a set of new visual techniques that changed the look of cinema. Directors previously working in television brought a preference for using telephoto zoom lenses and multi-camera setups to cover scenes from multiple angles simultaneously. Other filmmakers from low-budget and exploitation filmmaking favored shooting on location and using available light to keep the production costs of the films as low as possible. And those trained in documentary filmmaking introduced hand-held cinematography made possible by lightweight 16mm cameras. There are also equal accounts that point to the influence of the European New Waves—particularly the French *nouvelle vague* directors and Italian filmmakers such as Federico Fellini and Michelangelo Antonioni—on the visual experiments in new American cinema.[4]

While most of the standard histories of American cinema in the late 1960s and 1970s agree on the changes in visual aesthetic concomitant with the changes in the industry, the development of sound aesthetics in the 1960s is difficult to trace because the lines of creative exchange do not mirror the same patterns.[5] Indeed, the acoustic influences on American cinema of the 1960s did not all necessarily follow the same routes, and influences from popular music (rock music, singing stars, the synergy of theme songs), the recording industry (magnetic recording, multichannel mixing, "punch-in" editing), modern composition techniques (*musique concrète*, John Cage), and experimental cinema (nonsynchronous sound tracks, sound effect creation) all had strong influences on the development of sound aesthetics.

Unlike the purported influence of the French and Italian New Wave films on American cinema in the 1960s, the effect of New Wave filmmakers on the American film industry may not have been as direct. In fact, few directors from France or Italy started to come to the United States to make films until the end of the decade (Antonioni in 1969 with *Zabriskie Point*, Jacques Demy in 1969 with *The Model Shop*, Jacques Deray in 1972 with *The Outside Man*, and Jean-Luc Godard with his aborted *1 A.M.* [a.k.a. *One American Movie*] shot in 1968 with D. A. Pennebaker and Richard Leacock). Instead, a more direct point of influence came from the British New Wave directors of the 1960s who emerged from the Free Cinema movement started by Karel Reisz and Lindsay Anderson at the journal *Sequence*. Directors including Tony Richardson, Richard Lester, John Schlesinger, and John Boorman all got their start making fiction feature films that were highly influenced by a strain of British realism drawn from the Free Cinema documentaries and the "angry young men" playwrights—John Osborne, John Braine, Alan Sillitoe—many of whom provided the scenarios for the early British New Wave films.

Focusing on the lives of average, lower-class British citizens, the Free Cinema documentaries sought to apply the observational aesthetics drawn from Italian neorealism (location shooting, emphasis on lower class subjectivities, examining everyday activities) to a new form of documentary. Their maxim, "The image speaks, sound amplifies and comments," was expressed in the Free Cinema manifesto—signed by Richardson, Reisz, Anderson, and Lorenza Mazzetti in February 1956—and can be taken at face value when describing the films.[6] Because synchronous sound recording equipment was still very difficult to bring on location, especially for low-budget documentaries, the filmmakers relied on most of the images being shot without sync sound and the audio effects being recorded nonsync on portable tape recorders and added in post-production. As a result, they were able to experiment with how the manipulation of sound could augment and enhance the interpretation of the images for the audience. The earliest Free Cinema films, like Anderson's *O Dreamland* (1953) and Reitz and Richardson's *Momma Don't Allow* (1955), freely played with the juxtaposition of sounds and images without concern for absolute synchronization. Even by the time that sync sound became possible, the Free Cinema filmmakers still preferred to create an audio montage constructed from elements recorded on location and voiceover narration used to comment on the screen action.

"Up Front" Dialogue in the British New Wave

While each director involved in the British New Wave was strongly influenced by the visual and narrative style of other European directors—Lester by Godard and François Truffaut, Schlesinger by Alain Resnais and Truffaut, Boorman by Antonioni and Resnais—they collectively developed an approach to sound that is strongly associated with British cinema in the 1960s. Unlike classical American cinema, where there was an emphasis on close-miked dialogue and a spatial sense of realistic ambience, or the direct sound aesthetic of the French New Wave, the British directors developed an aesthetic of clean, almost airless dialogue that floated apart from background ambience and sound effects. Derived from re-recording lines in looping sessions, the British mixing strategies emphasized an "up-front" dialogue effect where the actors would respeak their lines of dialogue directly into the microphone on the dubbing stage. Even when the dialogue was offscreen or at a lesser volume, looping created a layered separation of the foregrounded voice and background hard effects and ambience.

In part, this foregrounding of dialogue was due to an unstated belief that the spoken quality of speech is as significant as its semantic function. Within British cinema there was a longstanding tradition of emphasizing accents, speech patterns, and stylistic usage as a way of commenting on the regional and class differences of the speakers. Beginning with the GPO documentaries, such as Arthur Elton and Edgar Anstley's *Housing Problems* (1935), and continuing in the Ealing Studio comedies and dramas, British cinema demonstrated a strong fascination with the accented voice as a marker of difference. Yet it was not until the British "kitchen sink" directors of the late 1950s and early 1960s that the value of the accented voice was fully exploited. With the adaptation of working-class plays like John Osbourne's *Look Back in Anger* (1956) and Alan Sillitoe's novels *Saturday Night, Sunday Morning* (1958) and *The Loneliness of the Long Distance Runner* (1959), local and regional accents were brought directly to mainstream audiences. As film historian Gene Phillips explained, British cinema "turned from the customary inbred theatrical dialogue that the screen inherited from the stage to a more naturalistic diction, allowing characters to speak in the language of ordinary people."[7]

When Tony Richardson was selected to work with Osbourne to adapt his play for the screen, he worked with Richard Burton to cultivate a Midlands accent to reflect his character Jimmy's lower-class upbringing in Derby. In the case of many subsequent films, casting choïces were based on the ability of the actors to fully embody their characters, especially through their character's voice. Albert Finney's Northern upbringing in Salford helped him with his Nottingham accent in Reisz's *Saturday Night, Sunday Morning* (1960), and Tom Courtenay's childhood in Hull was reflected in Billy's Yorkshire accent in Schlesinger's *Billy Liar* (1963). In addition to the centrality of the dialogue and the ability of accents to function as a form of referential shorthand, the nature of the voice itself came into emphasis. While the earlier British New Wave films like *Look Back in Anger* (Tony Richardson, 1959) and *Room at the Top* (Jack Clayton, 1959) featured the accented voice embedded in straightforward narrative storytelling, films like Richardson's *The Loneliness of the Long Distance Runner* (1962), Schlesinger's *Billy Liar*, and Anderson's *This Sporting Life* (1963) incorporated a strong sense of character subjectivity, often expressed through voiceover. In *Billy Liar*, the title character's flights of fancy are almost always introduced by his internal voiceovers, and in *This Sporting Life*, Frank Machin's (Richard Harris) flashbacks are inaugurated by having his teeth knocked out in a rugby game. In his stuporous state, Frank's confused mind becomes justification for a series of audio transitions that set up the story. As he is helped off the pitch, the roar of the crowd transitions into the sound of a rock drill, triggering a flashback to his time as a coal miner and introducing his relationship with Margaret Hamilton (Rachel Roberts). The majority of the narrative occurs through a series of flashbacks, and when during an emergency visit to a dentist Frank succumbs to the gas, his recollections are presaged by audio flashes as the sounds of the dentist's office merge with his memories.

Stimulated by the creative freedom expressed in Italian neorealism and the *cinéma vérité* documentaries emerging from France, the British New Wave directors preferred to shoot on location and emphasize the realism of the settings. While this approach translated into a strong sense of visual realism, the audio elements followed a different path. Because most dialogue was still being recorded as optical sound on film in both the UK and Hollywood, the camera and the bulky audio recording system had to operate in sync; this required a cabled "umbilical" connection that made it difficult to have a freely roaming camera and sync sound on location. In order to maintain synchronization between the camera and the portable recorder, both devices needed to be connected by an electrical cable that sent a sync pulse from one to the other. This severely limited camera movements when it was necessary to record synchronous sound. In the United States, the major studios went to great lengths to have mobile sound trucks and recorders to shoot synchronous sound on location. Yet British filmmakers were not as quick to latch onto the potential of using synchronous location recordings made at the same time as the image. In England, nearly all the sounds recorded on location were used only as cues and were replaced in editing. The dialogue was looped in post-production, recorded in a studio by actors timing their vocalizations to the lip movements projected on a screen before them. This resulted in a particular sound trope that cuts across most British cinema in the 1960s, and it was an aesthetic that was put to creative use by several New Wave directors. Because there was no "natural" link between sound and image, it freed the filmmakers to experiment with the expressive possibilities for sound in post-production. Instead of realism being tied to the synchronous recording of sound and image, new codes of representation emerged based around the dramatic needs of the narrative.

Consider the films of Richard Lester in which characters are heard regularly talking offscreen and the audience has to sift through the lines of their dialogue, tightly packed yet never overlapping, as they careen off one another. All the lines of dialogue, both offscreen and on, are mixed at approximately the same volume level despite their differing locations within diegetic space. For example, near the beginning of *The Knack . . . and How to Get It* (1965), Nancy Jones (Rita Tushingham) is riding into London on a city bus, surrounded by passengers near and far away, many of whom are chatting about her from all directions. The characters' comments should not be audible, spoken *sotto voce* or whispered at a distance from which she would not normally hear them, yet they are all heard directly because they are recorded at the same, clearly audible volume as the main dialogue. Richard Lester explained his approach: "I think that pictures, after all, create a mood. If you have to be specific, you can be specific. If you want to just create the mood, let's say, of a cocktail party, of snatches of dialogue overheard and indistinct comments of people in the street as they pass you by, it is perfectly legitimate that the audience also can't hear everything. In life, we seldom hear more than sentence fragments, and sometimes they are more interesting than hearing whole sentences."[8] Most notably this effect was first featured in Lester's *A Hard Day's Night* (1964) and was put to greater use in *The Knack*. In addition, the dialogue that we do hear in *A Hard Day's Night* is inflected by the Liverpool accents of The Beatles and Wilfred Brambell's Irish brogue. In fact, American producer Walter Shenson wrestled with the prospect of having the dialogue dubbed for American audiences (who by the early 1960s were still not accustomed to hearing British voices in popular cultural contexts), but after consulting with Lester he agreed to "let the audience work a little" to understand the scouse accents in addition to the very local references and slang.[9]

British audio technicians like Gerry Humphries, Stephen Dalby, Leslie Hodgson, Peter Handford, and Don Challis worked with the British New Wave directors to develop a sound aesthetic that best emphasized the local and regional dialects of the characters. As the sound aesthetics developed during the British New Wave made their way onto American screens, not all members of the audience were ready for this transition away from the simultaneity of voice and image. Film critic Manny Farber, for one, found the emphasis on offscreen dialogue to be distracting and noted in his 1966 essay "The Day of the Lesteroid" that "Lester's trademark is a kind of thickness of texture which he gets purely with technique, like the blurred, flattened, anonymous, engineering sounds which replace actors' voices."[10] Yet it was precisely this thickness of texture that Lester wanted to convey—the realism of the overheard conversation as filtered through the characters in the story, not the carefully regulated close-miked voice of American cinema—and the recording and mixing practices were ably suited to create the effect. In the close-miked technique, a character might mumble, whisper, reflect to herself, or otherwise speak very quietly, but we hear her with perfect distinctness and directness, as though she were standing next to us.

Another main characteristic of the British New Wave audio aesthetic is the nearly complete addition of sound effects and ambiences in post-production. A strong example of how this changes the aesthetics of the film can be found in the works of Joseph Losey, particularly in his 1967 film *Accident*. Although the complete replacement of sound could lead to some perceptual difficulties for the audience, in part because it was very difficult to match the spatial characteristics of the speaker's voice or sound effects to his distance from the camera, it provided an interesting freedom to the director, actors, editors, and soundmen. Indeed, the opening of Losey's film works as an audio zoom, bringing us, as well as Dirk Bogard's character Stephen, closer to the eponymous car crash that sets the story in motion. The fact that the film begins *in medias res* and much of the story is told in flashback is augmented by the way the sounds provide a dreamlike perspective on the events. When we hear offscreen sounds throughout the film, we are never entirely sure whether we are hearing the events synchronous with the image or if they are memories filtered through Stephen's consciousness. Instead of the gruesome details of the car crash that begins the film, Losey gives the audience small acoustic cues to how Stephen perceives the event by diminishing background ambient sounds and emphasizing the creak of a car door, the slosh of liquor in a bottle, the dripping sound of gasoline, all set against the distant chugging of a train and the trills of a song thrush just before dawn. For every "realistic" sound event in the film, there are arguably dozens of missing acoustic events that force the audience into a different perceptual relationship with the story. Thus the lack of spatial characteristics surrounding the closely miked voices is attributed to the perceptual activities of the characters rather than the limitations of the recording process.

Perhaps the most daring dissociation of sound and image occurs when, in flashback, we are shown the beginning of a tryst between Stephen and his former lover Francesca (Delphine Seyrig). In a condensed montage of images, the start of the affair is shown while snippets of banal dialogue are heard; yet at no point do the lines of dialogue match with the images. In fact, Losey goes so far as to make sure that the characters are not moving their mouths at all whenever the dialogue is heard, lending a further level of distanciation to the subjective flashback. Despite the banality of the dialogue, the asynchronism of sound and image prompts the viewer to question the meaning behind the lines and to realize that an affair is blossoming, a suspicion that is confirmed when we see the sly expressions on each character's face.

Losey's film shows a willingness on the part of British filmmakers to experiment with sound and image relations in order to augment a story and how it is being told. Losey's oneiric soundscapes, Lester's overheard dialogue, Schlesinger's conflation of dialogue and the internal voice, and the overarching emphasis on accented voices and regional dialects are just a few of the sound aesthetics that were developed by the British New Wave filmmakers. While these practices developed in parallel with the other New Waves around the globe, there was a direct connection between the British New Wave and American cinema when several of the directors came to the United States to make films for Hollywood. Starting with *The Loved One*, a parallel British invasion of the United States was occurring.

The Loved One (Tony Richardson, 1965)

The Loved One is a curious hybrid of British New Wave aesthetics and American cultural excess. Coming off the success of *Tom Jones* (1963), Richardson decided to concoct an adaptation of Evelyn Waugh's 1948 novel set in the hyperbolic epicenter of American culture, Los Angeles. Though the film was shot in California by cinematographer Haskell Wexler and produced by the American company Filmways Pictures, it was edited by British editor Tony Gibbs (assisted by Hal Ashby and Brian Smedley-Aston), and Richardson and Gibbs made the most of the post-synchronization of sound and image by carrying over the aesthetic practices developed in *Tom Jones* and *The Knack*.[11]

Importing the standard practice from the UK, the film's dialogue was almost entirely recorded in post-production, and there is a strangely hermetic quality to the film's soundscape that matches the cloistered Hollywood bungalows, smoke-filled studio offices, and airlessness of the Whispering Glades mortuary.[12] Indeed, there are moments in the film when sound effects are conspicuously absent (the interior cabin ambience of a jet or any sounds from a hovering helicopter), and there are several obvious mis-synchronizations of dialogue; yet rather than seeming like mistakes these augment the otherworldliness felt by the main character, recent British émigré Dennis Barlow (Robert Morse). In addition to its ethereal ambiences and sound effects, the film starts to break new ground for American cinema by demonstrating how sound and image can work independently.

Unlike Lester's films, *The Loved One* does not utilize much offscreen sound, nor do we hear conversations other than those shown on screen. However, there are several moments of sound-image disruption, the most obvious being visual cuts in time and space that conflict with the continuous conversations heard on the sound track. For example, when Dennis first meets with his Uncle Francis (John Gielgud) at the studio, the conversation carries on without a break even though their location jumps from the canteen to back lot in mid-sentence. This device is repeated later in the film and it creates an abrupt awareness of the manipulation of sound and image. The effect is one of distanciation that parallels Dennis's unease as a British man in the City of Angels and it plays to the film's comedic exaggerations. Although introduced in its rudimentary form in *The Loved One*, this technique comes to full effect in John Boorman's *Point Blank*.

Point Blank (John Boorman, 1967)

Although he is not often associated with the British New Wave after working extensively in Hollywood during the 1970s, John Boorman started as a director in television and documentaries before breaking into cinema. In the wake of The Beatles' success with *A Hard Day's Night*, Boorman was hired to direct *Having a Wild Weekend* (1965), a rock 'n' roll riposte to the Fab Four featuring The Dave Clark Five. Although the film was conceptually (and financially) inspired by Lester's film, Boorman chose to take things in a radically different direction. Instead of the wild, frenetic antics of the group, the film examined the growing boredom and disconnection of the younger generation in the UK through a picaresque narrative. In many ways it was the polar opposite to the brand-building effort of The Beatles' first film as it actively deconstructed The Dave Clark Five as pop culture icons. After its successful reception, Boorman was offered a contract to direct a feature film for MGM for which he received full creative control (i.e., final cut) over the picture. The resulting film, *Point Blank*, is one of the most fascinating works to come out of American studios in the 1960s. As Philip French pointed out, *Point Blank* is "an even more remarkable case than *Bonnie and Clyde* of the imaginative feedback into Hollywood of New Wave borrowings."[13]

Despite its surface story of a low-level criminal who is betrayed by his wife, Lynne (Sharon Acker), and his best friend, Reese (John Vernon), the film is more than just a simple revenge story: *Point Blank* takes as its target America's culture of success. The film centers on the aptly named criminal Walker (Lee Marvin), who is left for dead after being double-crossed and shot by Reese during an exchange of illicit materials in the decommissioned Alcatraz prison. In the film Boorman plays extensively with silence and sound effects to express Walker's confused state of consciousness through visual and audio dissociations. The fragmentation of the images—flashing back and forth between the present and the past—is accompanied by radical changes in ambience and reverberation as conversations are heard directly in the flashbacks and "replayed" in Walker's mind during the heist. In fact, as reviewer David Thompson noted, we are not even sure whether the events we see in the film are real or if they are subjective visions as Walker is dying in the rusting cellblock.[14] Walker's escape from Alcatraz is never seen and the film cuts from him stumbling and wounded to his return to the island, some years later, in an attempt to piece together what happened. Initially the film shows Walker, injured, wading into the San Francisco Bay while a female voiceover is heard explaining that no one has ever successfully escaped from the island prison. After a few moments we realize this sound advance is coming from a tourist boat that carries a healed and healthy Walker. To further complicate the narration and reveal the duality of the film's story, the tour guide's voice is heard in two parallel but slightly different versions: the direct sound of her narrating voice, signifying it as internal to Walker's mind, and a distorted version, presumably played over the boat's loudspeaker. Because we hear both versions, with the voiceover transitioning from the direct to the distorted, the effect casts into doubt whether the events we are seeing and hearing are real or imagined.

The abrupt visual transitions in time and space in the film, such as introducing flashbacks without context, are undoubtedly related to the experiments of the French *nouvelle vague*,[15] yet the sound aesthetics are much more in line with his fellow British directors from the 1960s. Boorman's use of offscreen conversations, disjunction between image and sound synchronization, sound advances, experiments with sound scale and reverberation, and post-sync dialogue looping were all techniques imported into Hollywood by the director. Ostensibly one of the boldest sequences in late 1960s American cinema occurs when Walker is tipped off to the location where his wife and Reese have been living in Los Angeles. It begins with an image of Walker striding down the corridor of the San Francisco airport with the sound of his reverberant footsteps heard on the sound track. As the film cuts between Walker and his wife going about her daily routine, the incessant cascade of his footsteps continues, reminding the viewer of the earlier sounds of Lynne's humming and Walker's reverberant footsteps in Alcatraz. Soon this sound motif begins to operate in counterpoint to the image as the unwavering sound of the footsteps continues over images of Walker arriving in Los Angeles, driving a car to Lynne's apartment, and following her up the external stairs (see figure 1). The sequence climaxes when the footstep motif ends, Walker bursts through the door, grabs Lynne, and fires six rounds into the empty bed where he had expected to find Reese. The build-up of Walker's explosive action is not constructed around either his dialogue or his actions (we only see the characters engaged in clichéd activities); rather, it occurs through the intensity and repetition of the footstep motif.

Indeed, the film is a catalogue of innovative sound techniques imported into American cinema. After he realizes that Reese is no longer living with his wife, Walker and Lynne carry on a "conversation" even though Walker does not utter a word. Sitting on the couch staring blankly into the middle distance, Walker remains stationary while Lynne responds to his unspoken questions. This one-sided dialogue serves to obfuscate the narrative drive of the scene and to bring

FIGURE 1. Walker (Lee Marvin) stakes out his wife's apartment in *Point Blank* (John Boorman, 1967) while the asynchronous sound of his inexorable footsteps is heard on the sound track.

into question whether the scene is fantasy or reality. Sound is used at several points to disrupt the viewer's direct perception of events by introducing abrupt transitions presaged by sound advances. Flashbacks of Walker's relationship with his wife are subtended by the sound of her humming, and the humming carries over to Walker waking up and finding her moved out of the apartment. Boorman is also open to the comedic potential of sound, as when Walker confronts the self-aggrandizing petty mobster Big John Stegman (Michael Strong) by posing as a customer at his used car dealership. Walker interrogates Stegman by systematically destroying the Chrysler Imperial he is test driving, repeatedly ramming it into highway support columns, as we hear the commercial for Big John's Used Cars playing on the car radio.

Like Lester's and Richardson's films, there is equal emphasis on how dialogue is heard in *Point Blank* and on its meaning. As Stephen Farber points out about the film, "There was sardonic detail in the dialogue too—in the unctuous language of a used car salesman, in casually overheard burblings of guests at a business meeting, in a criminal's concern about his swimming pool and his crabgrass."[16] Boorman's attention to sound's affective properties was augmented by his control over the film throughout all of its production and post-production. The film was one of the rare examples of the director's retaining final cut over the film, and MGM released the film as Boorman had conceived it.

This was not the case on Boorman's next film, *Hell in the Pacific* (1968), where independent producer Henry G. Saperstein retained final control and forced Boorman to include nondiegetic score music over the scenes of Lee Marvin's and Toshiro Mifune's World War II pilots lost on a Pacific island. As Boorman tells it, the minimalist and hermetic nature of the film unnerved the producers and they insisted on changing its soundscape:

> I emphasized their isolation by very discreet camera movements. So you've got a combination of no dialogue, just two men, very simple camerawork, and it becomes so simple that it almost doesn't exist. . . . The very strength of the picture is that it hasn't got any dialogue, or very little, and a powerful score would just let you off the hook, tell you what to think and feel.[17]

Even though he tried to convince the producers how a carefully planned and executed soundscape can help to convey the substance as well as the spirit of the story, Boorman had to wait until *Deliverance* (1972) before he could have the same level of control over his sound and image relations. Yet the attempt to curb the sound experiments in *Hell in the Pacific* did not mean that the British New Wave experiments had failed in Hollywood, for the aesthetics introduced by Richardson and Boorman were carried on by Lester in *Petulia*.

Petulia (Richard Lester, 1968)

Arguably the film that had the most influence on American cinema from a British New Wave director is Richard Lester's *Petulia*. The origins of the film are curious. After having worked with producer Walter Shenson on The Beatles' movies, American ex-pat Lester was asked by producer Raymond Wagner to return to the United States to helm the film. The source material, John Haase's novel *Me and the Arch Kook Petulia*, was published in 1966, and shortly thereafter Lester traveled to San Francisco during the "Summer of Love" to scout locations. What he found was the 1960s countercultural movement turning into something very different and much darker. Instead of the idealism that has been preserved by the media in the wake of such events as the Monterey Pop Festival and Woodstock, Lester perceived a growing sense of discontent that he captured in his film. The resulting adaptation drains away much of the "kooky" allure of the central character in lieu of an examination of American mores rusting away. Lester cast Julie Christie as the title character and George C. Scott as Dr. Archie Bollen, the object of her interest. The result is a powerful and pessimistic film that pushes the counterculture into the background in order to focus on the breakdown of the institution of marriage and the impossibility of living without consequences.

Petulia, a film dimly remembered today, was received well upon its release and was named in a poll of American film critics in 1979 as the third most important film of the prior decade.[18] Although retrospective critical evaluations noted the film's commercial success, few actually elaborated why the film resonated with American audiences. I argue it is because the stylistic devices, mostly imported from British cinema, meshed so well with the story of the alienated middle class and the impossibility for happiness in a commercialized world. These themes were best expressed in an audiovisual strategy that includes "near subliminal shots, time sequence out of order, and counterpointed sound."[19] Foremost among these techniques is a disjunction between sound and image that forces the audience into a viewing position where they become aware of the characters' emotional differences, one that distances them from an easy emotional participation with the narrative. Specifically, Lester and his editor Anthony Gibbs fragment time and space, giving viewers glimpses of both the past and the future as the story unfolds. The resulting "intercutting of one character's present with another character's past," as Stephen Farber observed, "gives us exactly the feeling that Lester must have wanted to convey, the sense of two lives being lived simultaneously, intersecting but essentially, perpetually disjointed."[20]

The flashbacks and flashforwards are very difficult to discern and follow on first viewing without noting the structural patterns that Lester and Gibbs

FIGURE 2. Richard Lester briefly introduces the musical counterculture of San Francisco with Big Brother & The Holding Company in *Petulia* (1968) before examining the decay of the upper middle class.

developed. In fact, the film starts with the juxtaposition between the image of an ambulance traveling through a tunnel, the siren only faintly filtering through the instrumental title music, and shots of the service area of a hotel as several guests in ball gowns and tuxedos are pushed in wheelchairs past the staff and laborers. The low-level sounds of the wheelchairs rattling in the hallway are all that is heard until a rapid burst of Big Brother & The Holding's Company's "Roadblock" interjects itself into the sequence, accompanied by close-up views of a guitarist's intense expression and the frenetic shake of a fringed skirt. The sequence continues to crosscut between the quiet electrical hum of the special guests being lifted in a service elevator and abruptly loud close-ups of Janis Joplin and the band in full swing (see figure 2). It isn't until two and a half minutes into the film, when a close-up on the sign "Shake for Highway Safety" is seen, that the viewer is able to connect the prior images and sounds. The ambulance delivering the well-dressed guests is linked to the ongoing dance, yet in Kuleshovian fashion they never occupy the same frame and the ambulance motif recurs throughout the film. In fact, the film progresses in such a way as to constantly throw into question the logic and timing of the events as they unfold, always placing the viewers into a position of narrative uncertainty unless they are able to discern the patterns surrounding the flashbacks, flashforwards, and "fantasy flashes."[21]

The way that Lester lets the audience distinguish these narrative fragments is though sound. The film functions much like a puzzle where solving the codes of cinematic construction is essential for understanding the progression of the story. Three main patterns of sound and image relation dictate the temporal structure of the film. During the present, there is a strong emphasis on the coincident presence of offscreen sounds and their effect on the visual images. In typical Lester fashion, much of the dialogue in the film is handled in subconversations—voices that are heard offscreen, often mediated through PA systems or other forms of communication, where the speakers are rarely seen or heard in sync with the image. As Lester explained this technique to Joseph Gelmis: "I think equal weight should be placed on dialogue, though not necessarily synchronized with the pictures. I would like to be able to move dialogue, so that it doesn't become just visualized theater. And so that words don't always have to come out of the mouths of the people who are on the screen."[22] This form of asynchronism frees up the expressive potential of the film and allows Lester to use the offscreen voices to direct our understanding of the onscreen images. Throughout the film, we hear unseen voices directing tourists in a Japanese garden, describing the topless dancers as well as the lunch specials at a gentlemen's club, and a "P.A. system in the hospital [that] calls—among the sick—for a doctor to move his parked car."[23]

Central to the offscreen voices are the background conversations that actively comment on the characters and their actions, often acting as surrogates for the audience and injecting doubt into the storyline. These background conversations are also interwoven with the ubiquitous voices from radios, televisions, loudspeakers, and PA systems to provide a steady bed of reproduced voices that regularly emphasize and reinforce the separation between the characters of Petulia and Archie. At nearly every interaction, we are made aware of their lack of privacy and how the mediated nature of modern society destroys intimacy. Different from most British New Wave films, nearly all the dialogue in *Petulia* was recorded on set and on location. The result is a sense of acoustic verisimilitude that anchors the characters in the diegesis even as the offscreen voices and sounds pull at the spatial anchorage of the onscreen voices. The spatially stable soundscapes in the sequences that unfold chronologically are contrasted by a more abstract soundscape for the flashbacks and flashforwards. All the flashforwards and fantasy flashes in the film, which can only be discerned after viewing the film in its entirely, are not accompanied by synchronous sound effects or dialogue. Generally they contain either nondiegetic music, which usually carries over from the preceding scene in the narrative present, or diegetic sounds of the present that bleed over from a prior location. The flashbacks, however, often have a slightly attenuated soundscape where several sounds are dropped out in favor of just one or two signature sounds. For example, in the first half of the film there are moments where Petulia's presence signals brief flashbacks to a car accident where a young Mexican boy is seen lying on the ground and later undergoing an operation where Archie is the lead surgeon. By the middle of the film the details of the accident are revealed in fragments via flashback, and as the images are truncated and slightly obscured, so too are the sounds that accompany them. Instead of seeing and hearing the accident directly, Lester conveys the details through an audio montage of a train crossing's bells, the rattle of a freight train, tire screeches, Petulia's anxious cry, and the echoey offscreen dialogue "Pull it off him!" as we see a car being lifted off the boy's bloodied leg. This evocatively minimalist soundscape allows the listeners to discern the difference among the present sequences and the flashbacks and flashforwards, helping them to navigate a complexly imbricated narrative structure.

Intriguingly, Lester used the sound and visual counterpoints in the film to construct a new model for American cinema in the late 1960s. As with his earlier films, the form of Lester's film was intrinsically tied to the cultural context and the audience's interpretation of the story details. As Stanley Kauffmann noted at the time of *Petulia*'s release, "The first flashes are bewildering, as if we stumbled on irrelevant privacies, but Lester is making a mental model for us: showing us how strangers think things unknown to us and how, as we get to know them, we know somewhat more of what they are thinking."[24] The sound and image relations in the film provided a template for future American films and had a direct influence on emerging filmmakers like Haskell Wexler and Jerry Schatzberg. Yet the closest relation to *Petulia* is from another British New Wave director, John Schlesinger.

Midnight Cowboy (John Schlesinger, 1969)

Like Lester, John Schlesinger experimented with sound and image relations in many of his earlier films before coming to the United States to make *Midnight Cowboy*. Indeed, his earliest features, *A Kind of Loving* (1962) and *Billy Liar*, serve as prototypes for several of the stylistic devices seen and heard in *Petulia* and *Midnight Cowboy*. Schlesinger's films generally represent a softening of the harder "kitchen sink" melodramas, and *Billy Liar* fused comedic elements with a harsh visual realism. It is precisely his mundane daily existence that forces Billy into his flights of fancy, and the encroachment of reality often pulls him out of his reveries. While Billy's fantasies are introduced by his voiceover, most of the transitions back to the present occur as audio intrusions into his idylls, as when the rhythmic knocking sound in Billy's wartime hero fantasy is revealed to be the sound of his mother rapping on his door to wake him. Schlesinger's next film, *Darling* (1965), also used a disjunction between sound and image to tell the story of fashion model Diana Scott as she relays the tale of her rising career to a reporter who is tape-recording their conversations. Even though Diana's descriptions of the events initiate flashbacks, her spoken description and the actual events often are not in accord; they are arranged in a separation that, as film scholar Gene Phillips comments, introduced "an ironic dimension to the story by having Diana narrate her life-story on the sound-track to a reporter from *The Ideal Woman* magazine."[25]

Borrowing from Schlesinger's own films as well as those of his compatriots, *Midnight Cowboy* stands as a summary of the audiovisual formal devices imported by the British New Wave directors into American cinema. The dimensions of the film's narrative follow the archetypal rags-to-riches story so cherished by classical Hollywood, yet Schlesinger put a sardonic spin on the tale by depicting ex-cowboy Joe Buck (Jon Voight) as a male prostitute trying to make it big in New York City. The distinguishing characteristics of the film include its complex layering of present-tense narration, Joe Buck's direct spoken commentary, the ubiquitous presence of radio and television voices offering extradiegetic commentary, and aural flashbacks that are not in sync with the image except during brief moments of synchresis.[26] Unlike their use in *Petulia*, the audio flashbacks in *Midnight Cowboy* make it rather disturbing and allusive for viewers, who are allowed to hear only the sound of the voice, not the other sounds that supposedly fit the scene. These disjointed audio flashbacks, used by Schlesinger before they became clichéd in the 1970s, blended Joe Buck's factual recollections of the past with his fantasies.

The trope of conflict between the fantasy of the audio and the reality of the visuals is put on display right from the start of the film. The sound of cowboys and Indians riding and shooting from an old Hollywood western are heard over the United Artists logo and a blank white screen. As the camera pulls back to reveal an abandoned drive-in theater, the nondiegetic sounds of its cinematic past are replaced with the diegetic sounds of a child riding on a creaking toy horse, the incessant wind, and Joe Buck's strongly reverberant voice singing, "Git along little doggies, for you know New York will be your new home." Through a form of audiovisual synecdoche, Schlesinger is able to encapsulate the power of the frontier myth as promulgated by Hollywood, the reality of its loss in late 1960s North Texas, and Joe Buck's narrative trajectory to New York, all in less than a minute of screen time. Over Joe's singing is heard a chorus of voices asking "Where is that Joe Buck?" along with flashes of a diner's dirty dishes and the agitated faces of its workers. Initially it is unclear whether Schlesinger is crosscutting the actual events or if they are Joe's fantasy flashes, but the latter is revealed to be the case when Joe's witty "responses" to the accusatorial voices, as he dresses in his newly purchased cowboy regalia, don't match their nonplussed manner when he tells them he is quitting. This transition between fantasy and reality occurs through an audio advance signaled by the inclusion of a slight amount of reverberation on the voices.

Also significant in the film is the use of Harry Nilsson's extradiegetic theme music, "Everybody's Talkin'," which accompanies the montage of Joe's journey to New York. Not only does it remind the audience of Joe's fantasy flashes that begin the film ("Everybody's talkin'") but it also functions as a reminder of Joe's delusions and his audio flashbacks ("the echoes of my mind"). Uniquely, the theme song fades out during the audio flashbacks, and the flashbacks usually end when a sound from the present creeps in and becomes recognizable. The technique is similar to the one Schlesinger used in *Billy Liar*, yet it takes on an added level of critical commentary as the flashbacks begin to change over the course of the film. What seem to be treasured recollections, such as Joe's courtship of his girlfriend "Crazy Legs" Annie or his grandmother softly singing "Mockingbird" to him, are revealed to be idealized versions of deeply repressed traumatic memories. With each iteration of the flashback, the bucolic patina peels away to reveal Joe witnessing the gang rape of Annie and his grandmother's borderline-incestuous licentiousness.

Alongside the use of the audio flashbacks, Schlesinger carefully built his soundscape out of narratively significant sound effects and ambiences. There are several moments in the film when narrative sound effects either trigger Joe's memories/fantasies or change the meaning of a scene. Whereas many of Joe's "memories" are accompanied by nondiegetic music and are intercut with the present, thus indistinguishable from reality, his dream sequence in Rizzo's (Dustin Hoffman) "apartment" takes on a disturbing acoustic effect. In addition to the dream images blanching from color to black and white, there are moments of red and white flashes accompanied by a disturbing *musique concrète* constructed of nonsynchronous reverberant twangs, electronic pulses, indistinct voices, sirens, and radio static. The chaotic sound montage matches the incongruity of the images—New York resident Rizzo appears alongside Joe's grandmother and Annie in Texas—and synchronization does not return until the moment when the radio static motif resolves itself into a local news broadcast and Joe finally wakes up. Similar effects accompany Joe and Rizzo's trip to The Factory as the sounds of the party and the music begin to distort and dissociate to reflect Joe's altered perception as he gradually slips into a drug trip.

Perhaps the film's signature sounds, and the strongest markers of its American origin, are the carefully cultivated accents and banter of the characters. These were accommodated through the use of portable tape recorders to record and replay rehearsals with the actors. Schlesinger explained how he and screenwriter Waldo Salt would incorporate elements of improvisation into the final version of the film:

> I like to rehearse a scene but not necessarily in the exact way in which it will be filmed. I always look upon the script as a blueprint that must be flexible enough to incorporate the things that develop once the production gets underway. We improvised certain scenes in *Cowboy* with a tape

> recorder running. . . . Then we took bits of what had been said by Dustin and Jon on the tape and put it into the dialogue.[27]

In addition, Jon Voight used a tape recorder to capture the voices of real Texans and studied the recording carefully to master his accent.[28] The result is a strong sense of realism drawn from the location shooting and the character stylizations of the actors, combined with the use of the sound track as a distanciation device, constantly forcing the audience to reassess the actions as statements of the characters. Indeed, the ubiquitous presence of media voices offers a running commentary on the actions in the film, as when the WAB radio news is accompanied by the visuals in Joe's fantasy about the "ideal man." This regular disjunction between Joe's fantasies and dreams becomes the motor for the drama within the film.

More than the stylistic tropes from French and Italian New Wave films, the techniques introduced by the British directors working in Hollywood would have profound and lasting influences on a new generation of American filmmakers. Richard Lester's use of asynchronous offscreen sounds and voices had a major influence on Robert Altman and return as meta-diegetic commentaries in his films in the 1970s. Lester's and Schlesinger's audio flashes were taken up by Haskell Wexler and Francis Ford Coppola as each director moved out of the studio and gained a larger measure of control, and sound and image disjunction and counterpoint were taken up by Jim McBride and Jerry Schatzberg in their respective first films. But perhaps the main difference between the British New Wave and their American counterparts was the preference for direct sound and location dialogue recording over looped dialogue in post-production. This technique led to a conflict in the codes of realism as they were expressed in fiction filmmaking, and that transition is best heard in the early independent films of director John Cassavetes.

3

TV and Documentary's Influence on Sound Aesthetics

The majority of the new directors who entered Hollywood in the 1960s were drawn from the ranks of television or documentary filmmaking. Filmmakers like Arthur Penn, Sidney Lumet, and John Frankenheimer brought techniques with them from television, such as multiple camera setups and wireless microphones, whereas others like John Cassavetes, Haskell Wexler, and Jim McBride borrowed the strategies of hand-held camera work and location sound shooting from documentary. These two strands of influence came into contact with the sound practices introduced by the British New Wave directors and hybridized into a range of sound aesthetics based around personal stylistic choices. Whereas the British aesthetic emphasized sound's narrative and emotional effect over realism, the American aesthetic placed an emphasis on both realism and sound's ability to render subjectivity. The waning years of the 1960s saw a move away from looped dialogue to the use of direct sound;[1] yet unlike the generation of filmmakers that preceded them, the new American directors sought to interrogate the ontological relationship between sound and image and to rewrite the cinematic codes of representation.

Codes of Realism and Their Changes

In the 1950s, with the move to widescreen cinema, color, and stereophonic sound, the cinematic codes of realism were moving in two divergent directions: one making a claim for greater perceptual realism, and the other valuing the attraction of greater spectacle. The fact that the technological advances of the 1950s came to be associated more with spectacle allowed the prior codes of black-and-white cinematography, the Academy ratio image, and monophonic sound to take on new meaning. With the rise of live television news reporting and newsreel cinematography, the codes of cinematic representation changed to include hand-held cinematography and location sound recording as techniques associated with a new realism.

Technology was a limiting factor when it came to the transfer of a documentary aesthetic into mainstream cinema. Until the 1950s, 35 mm cameras relied on mag-striped 35 mm sound recording film that ran on a sprocketed system of transport matched to the 24 frames-per-second speed of sound cinema. These recording devices were bulky and required that the sound recorder and the camera be tethered together so a sync signal could be sent from the camera to control the recording device. In 1951, Polish inventor Stefan Kudelski developed the first portable quarter-inch magnetic tape recorder, the Nagra, which he marketed through his company in Switzerland. Kudelski was also the first to develop a synchronization system where the motion picture camera's sync tone was recorded across the monophonic sound recording in two 180° out-of-phase bands that canceled each other out upon playback. The addition of this "Neo-Pilottone" system to the Nagra III tape recorder in 1958 made it possible for the device to operate in synchronization with the camera, thereby obviating the need for 35 mm mag stripe recorders on location. By 1962 third-party companies were offering crystal sync units that provided a regular sync pulse generated by the oscillations of a quartz crystal. When the camera and recorder were each fed a signal from their respective crystal sync units, it allowed them to operate independently, without a tether, and to stay in perfect synchronization with each other.[2]

The main aesthetic effect of lightweight recorders such as the Nagra III and crystal sync was the liberation of sound from the constraints of the image. No longer was the sound team literally tied to the camera on location shoots. Instead, sound recordists were free to experiment with microphone movement, recorder location, and a general separation of sound from the image. In the world of documentary, the ability for cameras and tape recorders to operate without the need for an umbilical cable allowed for direct cinema and *cinéma vérité* to develop as forms of observational cinema, where the camera and microphone could follow almost anywhere that their subject could go.[3] The result was a form of cinema based around reportage that commingled the objectivity of sound and image with an element of voyeurism.

Yet hidden behind the new documentary movements and their concurrent aesthetics was the continuation of an ontological fallacy—that the sounds heard in the film somehow were generated by the images and matched them perfectly.

One of the most important developments of separating visual recording from audio was that it pointed out the relatively arbitrary link between sound and image. The use of crystal sync allowed for the sound to be resynchronized perfectly to the image in such a way as to hide their separation during the recording process. While some filmmakers relied on the verisimilitude of synchronization to construct a sense of realism in their films, others exploited it to reveal the inherent disjunction between sound and image.

Independent filmmakers like John Cassavetes were the first to seize the opportunity to use low-cost, high-quality, quarter-inch magnetic recorders to record the sound and dialogue in the actual locations rather than relying on post-production looping to replace all the dialogue as he had on his first film, *Shadows* (1960). Because the Nagra III with external crystal sync was still a recent innovation when Cassavetes started production on *Faces* in 1965, he chose to use a Perfectone recorder in combination with two RCA lavalier mics and an ElectroVoice 642 shotgun mic to record the production dialogue and sound. Filmed on location in Los Angeles with a hand-held Éclair NPR 16 mm camera under a variety of conditions, the production recordings bear a strong stamp of realism due to the attendant sound of the spaces in which the scenes were filmed. However, the Perfectone recorder did not feature a crystal sync unit and despite the realism of the sound's spatial characteristics, Cassavetes discovered a major flaw in the system. According to Ray Carney:

> Cassavetes did not have enough money to transfer the sound until after the shoot was complete, and in the late summer of 1965, to his shock, he discovered that the soundtrack was unusable. The minor problem was that a lot of the film was poorly miked. . . . Even more disconcertingly, he discovered that the second-hand Perfectone that had been employed throughout the shoot was defective. It slowed down and gradually lost synch [*sic*] in the course of a reel. Cassavetes went to every sound lab in Los Angeles to ask for advice on how to resynch the sound. Everyone advised him that there was nothing to be done except throw the footage out and shoot it all again.[4]

Discouraged yet determined, Cassavetes wanted a sound aesthetic that preserved the live and improvisatory characteristics of the dialogue. "After having gone through what we had, it was out of the question for us to give up on the picture. So the next four months were spent cutting little pieces out of the sound, and adding to and subtracting from different places, trying to fit them to the action of the picture."[5] The months of editing required to resynchronize the dialogue with the images subsequently delayed the release of the film by three years, yet the resulting film makes a virtue out of a severe limitation by combining sounds and images in unique ways.

Because the director's preferred method of shooting long-take scenes was undermined by the drifting synchronization of the production recordings, he developed several strategies to provide a sense of spatial and temporal verisimilitude. First, in several places he shortened the shot lengths to an average of three to four seconds, cutting rapidly so as to retain sync between sound and image before it started to drift noticeably at around ten seconds. Second, if a shot ran longer than ten seconds either the dialogue would get trimmed to fit the images by cutting out any brief silences or the lines would be looped in post-production. In the former case this preserved the background ambience while creating the illusion of synchronization; in the latter it allowed dialogue to match lip movement but at the cost of a noticeably different background ambience. Even though looping was something the director wished to avoid, he recognized it was a necessary evil to complete the film. Third, and perhaps most significantly, Cassavetes discovered that he could manipulate the footage to hide the lack of proper synchronization. Very often he utilized visual masking to allow the dialogue from one take to be combined with images from another. If a character's mouth was obscured or if the character was off-camera, it became possible to use the character's voice from another take since the audience would not notice.

In combination, these techniques became the trademarks of Cassavetes's later films as he would regularly construct a highly realistic sound track out of elements recorded on location as well as dialogue and sound effects added in post-production. Even though he had started his career as a filmmaker who sought the absolute correlation between sound and image made possible through the documentary approach of live sound recording, Cassavetes rapidly discovered the aesthetic value in manipulating sounds to make them fit the emotions of his stories. When editing his next film, *Husbands* (1970), Cassavetes "manipulated the sound and dialogue, altering the meaning of old scenes and building entirely new ones through looping. For example, the conversation between [Peter] Falk and Noelle Kao on the bed was created by dubbing in new dialogue and using outtakes. Cassavetes cut most of the original dialogue out of the men's-room scene and added sounds of vomiting and farting (which were not in the original mix) to completely change its meaning."[6] Moreover, because the production recordings were considered a starting point for the final sound mix, Cassavetes regularly shouted out directions to his actors, knowing that he could edit out his interjections and add room tone from other takes to fill the gap. By the time of *Minnie and Moskowitz* (1971) and *A Woman under the Influence* (1974) he began to use the editing room as a place to change the overall meaning of the narrative through manipulation of sound.[7] These techniques, while not exclusive to Cassavetes, marked a radical change in the way that film sound was being conceived in the 1960s. Emerging filmmakers Jim McBride and

Jerry Schatzberg, as well as established directors Arthur Penn and Haskell Wexler, all sought to use a combination of location sound recording and post-production manipulation to create sound tracks that combined the verisimilitude of documentary aesthetics with the dramatic needs of fiction storytelling.

David Holzman's Diary (Jim McBride, 1967)

Independent filmmaker Jim McBride was one of the first directors to experiment with the narrative and stylistic possibilities of revealing the separation of sound and image made possible by lightweight documentary film equipment. His debut film, *David Holzman's Diary*,[8] is putatively the autobiographical record of a filmmaker trying to record the details of his life before he has to report for the draft during the Vietnam War. The film begins with Holzman, using an Éclair NPR 16 mm camera, shooting an image of himself in a storefront window being recorded by a video camera. Over this is heard the narration, "Test, test—OK, this is the . . . this is the story, this is very important, this is a fairy tale—please pay attention," and then the image cuts to a shot of Holzman filming himself in the mirror of his apartment. The oblique nature of the voiceover in relation to separate images inaugurates a pattern of pointing out the disjunction between sound and image and how their combination is ultimately controlled by the filmmaker. Holzman's story begins when he is seen in front of his camera, literally bringing the image into focus, narrating the events of his life from the living room of his apartment. In the background a mirror reflects the Éclair camera. Holzman is seated, wearing a lavalier microphone around his neck with the Nagra III recorder on the table next to him (see figure 3). As he adjusts the sound recording level, after previously shifting the camera focus, the audience is made aware of the recording apparatuses and their separation in space during the filming process. If this is not evident enough, after a fumbling start Holzman steps up to the camera and turns it off, while the sound of him moving around the apartment continues over black leader. After several seconds he starts the camera again and continues telling his story.

This effect of breaking the continuity of image or sound occurs several times. The film functions less as a coherent integration of sounds and images to construct a transparent narrative than as a series of audio or visual dispatches from the time the film was made. As Peter Hogue pointed out in a retrospective review, "The film evokes the anxious summer of 1967 (most explicitly, via faintly overheard news reports of the Vietnam War, ghetto riots, et al.)."[9] Holzman's audiovisual diary uses a combination of directly recorded sounds, retrospectively recorded voiceovers, and recordings of radio and television broadcasts to construct a soundscape of New York in the summer of 1967. Whenever he is seen in front of the camera, he is usually accompanied by the

FIGURE 3. *David Holzman's Diary* (Jim McBride, 1967) displays its documentary aesthetic by pointing out the Nagra III recorder and lavalier microphone used to record the sound.

lavalier mic and Nagra tape recorder to reinforce the simultaneity of sound and image. To further add to the verisimilitude of the story, each day of recording is date-stamped by Holzman to make the audience aware of the passage of time over a one-week period from 14 to 20 June. Many of the audio recordings bear an additional hallmark of realism due to the way in which the single omnidirectional lavalier mic picks up Holzman's voice as well as extraneous noises: breathing, handling noises, room reverberation, passing traffic, and radio broadcasts that filter in from nearby rooms. Yet it is precisely the nature of these "accidental" sounds that establishes the realism of the film, and the viewer is constantly made aware of how sound can be applied to image in a way that reinforces its ontological realism.

Holzman's diary is not presented as just a first-person narration of the events occurring around him. Instead it intercuts his auto-interviews with interviews of friends and strangers, voyeuristic images of his girlfriend Penny and his neighbor "Sandra," and images and sounds captured on the streets of New York. Unlike the auto-interviews, these other passages often feature a counterpoint between the images and the sounds placed over them. Perhaps the purest audiovisual counterpoints occur when Holzman chooses to combine images shot on the Éclair with sound recorded separately on the Nagra. At several points in the film we are shown images of New York residents on subway trains, on park benches, and walking down avenues, each subtended by sounds that bear no direct relationship to the images. For example, early in the film, slow motion images of the denizens of West 71st Street are accompanied by a WABC radio news report, and it is clear that the late-night time of the broadcast, announcing the names of those killed in Newark race riots, contrasts with the bright daylight images. Later, as the camera rakes over dozens of pensioners spending their morning on rows of sunny park benches, the sound track replays the transcript of a vote tally from the United Nations. At other points there are sequences where the image stops while the sound carries on, foregrounding the relatively arbitrary relationship between sound and image. Holzman even acknowledges the limitations of his recording apparatuses in his narration, saying, "At about this point the tape ran out." These conflicts between sound and image not only point out the constructed nature of the film but also underline the way in which cinema reconstructs elements from different times and places to create the diegesis.

The film also uses images shot without sync sound and relies on narration to explain them to the viewer. This first occurs when Holzman secretly films his girlfriend as she sleeps. The sequence is introduced by Holzman announcing, "About two minutes ago, Penny walked out of here," over a black image to construct narrative suspense. His narration continues to explain how he was struck by the image of Penny being like an exhibit at a museum, "where everything is just . . . perfect," at which point images of Penny asleep in the nude appear as Holzman's voiceover explains that he chose to film her surreptitiously despite her earlier protestations. As the sequence continues, the voiceover ends when Penny wakes up to angrily confront Holzman and his camera. Even though she is seen shouting and hitting him, the audience hears only the static hum of tape hiss and no other diegetic sounds are heard. Later in the film voiceover is used to clarify the meaning of the images, such as when Holzman is confronted by a police officer for spying on Penny in her apartment. Following nearly three minutes of a silent hand-held tracking shot peeking into numerous windows, Holzman's voice returns to explain that the footage is about to be interrupted. "Watch him," he says in reference to the police officer, "he's going to hit me now," as the officer steps up to the camera and the screen goes black.

Even though the film starts out as pure reportage, a document of Holzman's world, as Holzman's life breaks down it is mirrored by the structural breakdown of the film. His auto-interviews continue, yet there are moments when Holzman is at a loss for words, remaining mute until the film in the camera or the tape in the recorder runs out. The film itself comes to an end when the impulse to document his life can no longer continue and what we see and hear are constructed out of the disparate sources available to him. In the last sequence, a scratchy, attenuated version of Holzman's voice is heard explaining how he is recording the monologue in a Duodisc recording booth and that he will photograph himself in an adjacent photo stall. As the dialogue continues, still photos of him holding the Duodisc appear on screen as he explains that his apartment was burglarized and the Éclair camera and Nagra recorder were stolen. This last, irreparable caesura of sound and image also signals the end of his project. Wishing that he had learned something from his audiovisual analysis of his life, Holzman concludes, "like Bartleby the Scrivener, 'I would have preferred not to have done this,' but I did it," as the disc recording ends and the final still image fades out.

Only after the end of the film and the title credits is it revealed that David Holzman was a fabrication of the director Jim McBride and the actor/writer L. M. Kit Carson, who also played Holzman. With this revelation it is possible to see how the filmmakers relied on audience understanding of the audiovisual codes of representation subtending documentaries to allow themselves to be tricked into believing the fiction and retrospectively assess the fictional construction of David Holzman. McBride continued to experiment with the use of direct sound aesthetics in his subsequent films, and his post-apocalyptic love story, *Glen and Randa* (1971), used location sound recordings only and had no score. McBride was one of several directors who were searching for ways to use sound to reinforce realism while also conveying a sense of objective expressiveness. At the same time that McBride was foregrounding the ontological realism of location sound recording in *David Holzman's Diary*, Arthur Penn was combining location sound recording with a heightened, affective approach to sound effects use in the breakthrough film of the new American cinema, *Bonnie and Clyde*.

Bonnie and Clyde (Arthur Penn, 1967)

The project of *Bonnie and Clyde* originated outside of the studios with a script from *Esquire* magazine writers Robert Benton and David Newman. Benton and Newman crafted their tale by borrowing the narrative details from John Toland's 1963 book *The Dillinger Days*, a chronicle of the early 1930s exploits of gangster John Dillinger and filled out with other accounts of Babyface Nelson, Machine-Gun Kelly, the Barker Gang, and the Barrow Gang, fronted by Bonnie Parker and Clyde Barrow. Benton and Newman relied on *The Dillinger Days* as an authentic source for their history, but it is clear that Bonnie and Clyde were only a minor focus of the book. Toland's portrayal of the couple leaves large gaps in their history and drops in wide, unsubstantiated speculations about Bonnie's sexual appetite and Clyde's sexual orientation. It does, however, form the central narrative trajectory of Benton and Newman's script from Clyde meeting Bonnie in Dallas through their ultimate ambush by ex-Texas Ranger Frank Hamer. However, several things from Toland's book were left out, specifically Bonnie and Clyde's utter brutality to most people they encountered, even those who tried to help them.

Why were these changes made to Toland's original history? Partly because Benton and Newman were enamored with the French *nouvelle vague* and they were writing the script with an eye toward François Truffaut directing. They modified their story to bring it more in line with the "lovers on the run" plot line expressed in numerous film noirs, especially with Jean-Paul Belmondo's portrayal of Michel Poiccard from Jean-Luc Godard's *À bout de souffle* (1960) in mind.[10] When Truffaut turned down the offer to direct because of his commitments to *Fahrenheit 451* (1966), the script was shopped to Godard, who also passed to make his own gangster film with Belmondo, *Pierrot le fou* (1965). Despite these initial setbacks, Truffaut mentioned the script to Warren Beatty while the actor was in France during the pre-production of *What's New Pussycat?* (1965).[11] Impressed, Beatty bought the rights to make the film figuring himself as producer and the young Bob Dylan, who bore a close resemblance to Barrow, as its lead. Beatty took the film to Warner Bros., where he pleaded with Jack Warner to provide financing, and although Warner himself turned down Beatty the studio's production head, Walter MacEwen, offered to fund the project if Beatty would take on the title role and a 40 percent producer's cut in lieu of his regular actor's salary.[12]

Arthur Penn was brought in to direct by Beatty because he had worked with Penn on *Mickey One* in 1965. After Anne Bancroft and Patty Duke both won Academy Awards under his direction in *The Miracle Worker* (1962), Columbia Studios gave Penn complete creative control over *Mickey One*. Penn's stylistic experiments in the film, shot on location in Chicago, mirror the use of post-production dialogue replacement by the British New Wave directors while also showing an awareness of Godard's jump cuts and narrative fragmentation. Even though the film was a financial bust, it represented a new style of filmmaking that had hitherto been unseen in the Hollywood studio system. *Mickey One* was highly personal and, according to Penn, it explored the possibility of telling a story on a "symbolic and metaphorical level" instead of being purely narrative.[13] Although it garnered Penn critical praise for his liberal use of New Wave aesthetics, the film's style distanced audiences and it was a commercial disaster.

The film that followed, *The Chase* (1966), was produced by Sam Spiegel for Columbia and represented a major attempt at creating a star-studded box-office draw. Not surprisingly, Penn's vision of the story, roughly a tale of southern racial intolerance and the impossibility of an honest sheriff to control a town's inclination to mob justice, was not the type of film usually made by a studio like Columbia. Although Penn's control of the visual medium is quite impressive, the editorial process was supervised by Spiegel while Penn was rehearsing the theatrical debut of *Wait until Dark* on Broadway.[14] As a result, the emotional impact of the film was blunted by Spiegel's final cut and Penn gained a great mistrust of working in the Hollywood studio system.

But when Beatty approached Penn to direct *Bonnie and Clyde*, the actor's secondary role as producer provided the needed buffer between the director and Warner Bros. After his negative experience with Spiegel, Penn determined that *Bonnie and Clyde* would be made in the same fashion as *Mickey One*, with only a limited number of scenes shot on the studio lot. A salient feature of Penn's directorial style was his commitment to producing the film using the preferred New York method of small-scale production teams and location shooting. This gave Penn control over the production schedule as the crew traveled throughout Texas in the winter of 1966–67, providing the film with realistic location cinematography. In addition, by doing so Penn was able to fulfill his desire to cast real-looking people for all extra roles by hiring locals as they moved from site to site. While the closeness of the production crew allowed Penn a large measure of freedom when shooting the film, he also developed a strategy that gave him relative autonomy in the post-production process. Instead of editing and mixing on the Warner Bros. lot in Burbank, Penn assembled his own post-production team and edited the film in New York.

His principal ally and collaborator was film editor Dede Allen. Allen had worked in both New York and Hollywood as a picture editor since the mid-1940s, and she started in the film industry working as a sound effects editor at Columbia Studios.[15] During that time she developed an intimate understanding of the complex relationships between sound and image, an awareness that manifested itself in the careful attention paid to sound in all of her subsequent work. After returning to New York in the early 1950s, Allen started cutting picture for industrial films before she moved into feature film editing. Thanks to a friendship with former sound editor Robert Wise, Allen's major break came when he asked her to edit his 1959 picture *Odds against Tomorrow*. Wise chose Allen because she came out of sound editing and was able to understand how Wise wanted the sound track to work in his film.[16] As a result, Allen was able to cultivate a method for cutting the film to meet the narrative demands of the picture while consistently strengthening the images with an astute sense of potential sound usage. In particular, Allen was conscious of the powerful effect of sound bridges and she would regularly "bounce sound forward or backward to propel something [in the story]."[17] According to Paul Monaco: "Dede Allen always worked closely with the film's post-production sound people, coordinating her own artistic choices in the editing with the spots in a film where sound was being crafted in new ways. Dialogue lines might be laid on top of one another in the final editing, or strung together without pauses that had actually been recorded during a take."[18] Moreover, Allen utilized an idiosyncratic system devised by cobbling together two Moviola editing machines to compare different lines and takes in a scene. Even before the advent of editing tables, she was able to cut picture and sound simultaneously instead of simply editing the image track without a corresponding sound track.[19] Her unique style developed because of the encouragement of her directors and the relative openness of the New York system.

Working with New York–based directors Robert Wise, Robert Rossen, Elia Kazan, and Arthur Penn, Allen had an unprecedented amount of freedom and flexibility in developing her own approach to the editing process. Specifically, it was under her tutelage that a number of editors were able to train and develop a "New York school of editing." When asked about this, Allen replied:

> I call it the Arthur Penn school of editing. For years we were lucky enough to be able to train a lot of people in New York. Directors like Arthur Penn came along and allowed us to have more people in the cutting room because he shot more film. It was a wonderful educational school for Stephen Rotter, Richie Marks, Jerry Greenburg, and many other people who came through the Arthur Penn cutting rooms. [The 1960s] was a very rich period for New York. There was a vitality and an independence in being away from the studios.[20]

Allen's work on Penn's film was different from her work for other directors due to Penn's belief in filmmaking as a collaborative art. In order to make the film as effectively and efficiently as possible, Penn envisioned the production process as a democratic endeavor rather than as a rigid hierarchical structure. Rather than simply asserting a large measure of authorial control over the production, Penn preferred to choose his collaborators carefully, readily acknowledging each of their strengths in the process.

In the case of *Bonnie and Clyde*, even though Penn had arrived late on the project, the final film bears the stamp of his presence and control, a control that was achieved by the careful selection of the film's crew. In order to craft a clearer story line, Penn hired his friend Robert Towne to tighten up the script and reorder the sequences. Specifically, Towne modified the script to emphasize the eruptions of violence throughout the film.[21] Penn also brought on costume designer Theodora Van Runkle and set designer Dean Tavoularis to evoke a familiar sense of the 1930s by using the photography of Dorothea Lange and Walker Evans along with the paintings of Andrew Wyeth and Maynard Dixon as sources. Finally, musician Charles Strouse was hired to add a nontraditional sound track that blended the bluegrass music of Flatt & Scruggs with more subtle non-orchestral background music. But the most valuable member of Penn's team was Dede Allen.

Editing the film in New York not only allowed Penn to work at a more leisurely pace directly with Allen, but it also freed him from the constant scrutiny of the studio. Although he worked closely with Allen during the editing process, Penn allowed the editor's own sensibilities to shape the progress of the film. According to Allen, "Arthur Penn is a very erudite man. He can be almost frighteningly articulate, but he never tells you [what he wants], he lets you find it."[22] To achieve this, Allen devised a unique method, bringing both sound and picture editing together in the editorial process:

> I work with a sound editor. It's a conceptual thing; we have long discussions. During the rough cut, I only deal with one track and if I choose to, I premix some things. I do a lot of intercutting of music and sound effects and things. If it's playing, I leave it alone on one track, because premixing costs money. But if it isn't playing, I go ahead and do a premix. That's the key to anything. If it's not playing, that is a problem I must solve before I show it to a director for the first time. For the purpose of showing a scene to a director I might also do what I call a "quick and dirty," which is a premix done on two dubbers. You can't even do that in Hollywood because of the union thing and in New York I guess you shouldn't, but we do.[23]

Allen was not afraid to break with the strictly regulated union practices and to incorporate sound editing into the overall picture editing procedure, a process that allowed her to conceive of the film in both its audio and visual components at all times.

In distinct contrast to the British use of post-production dialogue and effects recording, *Bonnie and Clyde* featured a large amount of production dialogue. This was in keeping with an East Coast resistance toward looping if it was possible to use production tracks, but the rationale had less to do with expediting the production and more with emphasizing the verisimilitude of the sound and image relations. The film was meant to evoke both the black-and-white Depression-era photos taken by WPA photographers and earlier gangster romance films,[24] but the color cinematography was an element that could potentially distract audiences from narrative concerns. To play against this effect, the sounds in the film were marshaled to express the dramatic weight of the sudden episodes of violence. Allen's careful placement of sound effects and dialogue within the film allowed for an aesthetic linkage to the underlying violence present in 1930s American society. To achieve this she used increases in volume to emphasize the impact of a cut, especially in relation to the sudden paroxysms of violence in *Bonnie and Clyde*.[25] In her work with Penn, Allen developed this method to shock the audience, which had previously been lulled into a secure listening pattern by the steady flow of dialogue and music. The hypostatized violence of the sound effects not only reflects the violence in the diegesis, but also serves as a sharp reminder of the ongoing war occurring in Vietnam. As Robin Wood reminded us, "Penn brings us physically so close to his [characters] that it is difficult to remain detached," and in doing so, we are even more shocked when these characters meet with violent ends.[26]

The escalation of violence is met with an elevation of both sound track volume and the visceral quality of the gunshots and other acoustic effects. The extreme loudness of the first gunshots, heard when Clyde is showing Bonnie (Faye Dunaway) how to shoot, undercuts the light atmosphere of the narrative moment (see figure 4). This undercurrent of acoustic violence reaches the surface when during a robbery Clyde is forced to shoot a bank employee who tries to stop them from escaping. The peripeteia of the moment is enforced by Allen cutting from the sound of the gunshot to the fractured car window with the banker's bloodied face behind it. A closer listen reveals not only Allen's subtle acceleration in the pace of the shots within the sequence leading up to the killing, but the sound track building to a furious crescendo with the grinding gears of C. W. Moss's (Michael J. Pollard) getaway car and the sound of the bank alarm. In combination, the sequence moves from Bonnie and Clyde's lighthearted cajoling of the bank customers to a moment of sheer horror as the audience is placed in the point of view and point of audition of both Clyde and the bank clerk being shot.

FIGURE 4. Editor Dede Allen used the extreme loudness of the gunshots to cultivate an undercurrent of violence during the earlier comedic scenes in *Bonnie and Clyde* (Arthur Penn, 1967).

Penn rejected the idea of a traditional score in favor of using Flatt & Scruggs's contemporaneous "Foggy Mountain Breakdown" to evoke the period and a mood, often in counterpoint to the narrative weight of the actions being depicted on screen. The banjo music also provides a frenetic pace to the scenes, but one that is often undercut by moments of abrupt and excessive violence, thereby shocking the viewer while also contradicting any spectatorial security provided by the upbeat music. Most of the other music in the film was supplied by bluegrass artist Charles Strouse as well as an introductory song by Rudy Vallee that subtends the opening credits. In an era of fully scored films that often filled the film with music from beginning to end, Penn's sparingly evocative sound track stood out. Indeed, the fumbling first sexual encounters between Bonnie and Clyde are depicted without any music, making them extremely uncomfortable for the audience while matching the conflicted emotions of the characters quite well.

When the film opened nationwide in September 1967, the first round of reviews were unequivocally negative. It wasn't until Pauline Kael's *New Yorker* review that the critical tide started to turn. In her review, Kael identified that the film tapped into a palpable current of emotion that ran through the youth of America in the late 1960s. By making us follow the story so closely from the perspective of real criminals, "*Bonnie and Clyde* keeps the audience in a kind of eager, nervous imbalance—holds our attention by throwing our disbelief back in our face."[27] Perhaps the most salient evidence of the turn in public opinion was that after its dismal initial run the film was re-released to massive success in the spring of 1968, a feat engineered by Beatty when Warner Bros. was ready to shelve the film. Furthermore, Theodora Van Runkle's costume styles started a retro fashion trend and the bluegrass music of the sound track caused a massive demand for a then nonexistent sound track album.

In contrast to the film industry and its sluggish response to the youth market in the 1960s, the recording industry was far faster to react to current trends. Within a month of the film's re-release a number of albums and songs about Bonnie and Clyde were flooding the marketplace. In addition to a vogue for bluegrass music, the characters of Bonnie and Clyde had gained a certain cultural cachet that was being exploited in numerous popular songs. The characters were immortalized in songs by Mel Tormé, Serge Gainsbourg and Brigitte Bardot, Merle Haggard, and Georgie Fame as they circulated in the popular memory of music listeners around the world. In particular, the songs sought to emulate, and even replicate, the heightened acoustic aesthetics of the film in their musical structures. While Gainsbourg and Bardot's duet contains a translation of the poem "The Ballad of Bonnie and Clyde," written by the real Bonnie Parker and used in the film, Georgie Fame's "The Ballad of Bonnie and Clyde" included the sound of machine-gun fire, which led to the disc being banned in several countries.[28] In early March 1968, Warner Bros. records finally released a sound track album that compiled both the music from the film and a dozen dialogue excerpts. This odd pairing of dialogue with the music from the film makes the sound track less a commercial music property and more of a souvenir for the filmgoers interested in reexperiencing the visceral effect of the film. The dialogue let the listeners literally replay the movie in their own homes, thereby reliving the experience through the sounds of the film.[29]

The success of *Bonnie and Clyde*, both as a commercial studio film and as a cultural artifact, had a profound effect on Arthur Penn and his filmmaking procedure. He continued to work with Dede Allen on most of his subsequent pictures, and his next film, *Alice's Restaurant* (1969), reversed the commercial process by taking a popular musical recording and using it as the inspiration for the film. Arlo Guthrie's "Alice's Restaurant Massacree" (1967) is a heavily narrative song that served as the basis for the story, and by expanding the song into a feature-length film Penn sought to forge a direct connection between Hollywood filmmaking, popular music, and the growing youth protest movement in America. Although the film was not entirely successful, in part due to the glut of other "youthquake" films that appeared in the wake of *Bonnie and Clyde*'s success, it did mark a continued commitment on Penn's part to a democratic form of collective filmmaking.[30] Moreover, it was one of the first films made within the Hollywood system to embrace the "New Left" and to try and articulate its political imperatives on the level of film form. Although this was relatively rare within mainstream filmmaking, one other film from the late 1960s managed to achieve this while also creating a dialectical relationship between its sound and image tracks: *Medium Cool*.

Medium Cool (Haskell Wexler, 1969)

When considering the great American films of the late 1960s, it is often easy to skip over *Medium Cool*. Lost amid a flurry of youth-oriented films, *Medium Cool* often gets written off as a curiosity: a feature-length fiction film directed by cinematographer and documentarian Haskell Wexler. Best known for his lens work for Mike Nichols (*Who's Afraid of Virginia Woolf?* [1966]) and Norman Jewison (*The Thomas Crown Affair* [1968]), in 1968 Wexler was given the chance to direct his first fiction film for Paramount Pictures. The result met with the ire of both the newly formed MPAA ratings board and the executives at Paramount, who found its language and content to be extremely offensive and nearly did not release the film. The critics generally dismissed *Medium Cool* as a cheap attempt at political persuasion wrapped in a patina of art cinema effects culled from European New Waves. While this was a growing trend among American filmmakers, Wexler demonstrated a deeper understanding of the mechanisms of cinema as he tried to make the audience aware of the audiovisual illusion of the onscreen fiction. In what is perhaps the most Brechtian film ever made in Hollywood, *Medium Cool* places itself in the rubric of *cinéma vérité* only to disassemble and question the truth value of film itself.

Although Wexler is a unique figure in cinematography, having cut his teeth shooting both documentaries and commercials before moving into feature films, like Arthur Penn he is also associated with the New York school of filmmaking. Unlike most directors who switched to feature films after starting in television or documentary, Wexler continued to make, and still continues to make, documentaries. For Wexler, documentaries were more than just a route into feature films: they represented the mark of political and social commitment, a commitment that is fully evident in *Medium Cool*. In addition to his work on a number of politically charged documentaries in the 1960s, especially *The Bus* (1965), his feature-length film about war protesters on their way from San Francisco to Washington, D.C., Wexler also worked with several influential new fiction directors. Starting as director of photography on three films for former documentarian Irvin Kershner, Wexler shot *America, America* (Elia Kazan, 1963), *The Best Man* (Franklin J. Schaffner, 1964), *The Loved One* (Tony Richardson, 1965), and *In the Heat of the Night* (Norman Jewison, 1967). Most importantly, his Oscar-winning job on Mike Nichols's *Who's Afraid of Virginia Woolf?* made it possible for him to work with H&J productions to fund his long-awaited feature filmmaking project as director. *Medium Cool* represents a combination of Wexler's documentary sensibilities, especially the candid use of direct interview *cinéma vérité* camera techniques and hand-held framing, with a healthy influence from the British New Wave absorbed through his work with Richardson. The result is an intimate film that fractures many assumptions of narrative stability to force the viewer into a critical relationship with the fictional material on both the image and sound tracks.

This collision of styles manifests itself from the start of the film as the audience accompanies a Chicago television camera crew as they film and record a car accident in the pre-dawn light. Instead of beginning the film with a traditional title sequence and score music, Wexler confronts the audience with moral and narrative ambiguity as the familiar Paramount logo (with its recent Gulf + Western parent company listed beneath)[31] fades to reveal the accident as we suddenly hear the sound of a car horn fill the sound track. After the television sound recordist disconnects the stuck horn, the moans of the injured driver are all that are heard as the reporters continue to film the accident with callous disregard for the victim. Like *David Holzman's Diary*, the separation of sound and image is made evident as the sound recordist and cameraman ply their recording devices over the wreck as the driver cries in distress (see figure 5). The minimal sound and image information of the opening scene is abruptly

FIGURE 5. Haskell Wexler's *Medium Cool* (1969) begins with the grating sound of a car horn as television cameraman John Cassellis (Robert Forster) and his sound recordist Gus (Peter Bonerz) film a car crash in Chicago.

undercut by Mike Bloomfield's driving musical score that accompanies the delivery of the accident footage to the studio by motorcycle messenger. Shot from the back of the motorcycle using a seat attachment and a camera with an extreme wide-angle lens, the brief sequence flows effortlessly underneath Bloomfield's twangy guitar and martial drumbeats. The abrupt shift from one aesthetic regime to another sets the tone for the rest of the film and any familiar spectatorial position is undercut when the audience notices that several of the fictional elements are shot with a hand-held camera while the "documentary" sequence of the motorcycle's journey through the city is presented in a highly stylized form. Moreover, the shock of the first sequence is redoubled when we discover that the seemingly ghoulish reporters are, in fact, the central characters in the film.

In this way Wexler forces the audience to question its perceptions of both the images and sounds they encounter and to weigh the truth value of each. This dialectical approach to sound and image construction is unique for a Hollywood film. Borrowing heavily from the fragmentary narrative constructions of the British New Wave, most directly from *Petulia* (Richard Lester, 1968), Wexler sought to create a film that fused his specific interest in social commentary with the new formal experiments coming from Europe. His goal was to combine *cinéma vérité* and fiction film approaches, and as he explained, "I have very strong opinions about us and the world and I don't know how in hell to put them all in one basket."[32] To achieve this, Wexler confronts the audience with a disjunction between fictional narrative and documentary "events" in the film.

Originally planned as an adaptation of Jack Couffer's novel *The Concrete Wilderness* (1967), about the flight of thousands of Appalachians to Chicago in search of work, Wexler decided to interweave Couffer's story with actual location shooting in the north side ghettos. In addition, the film is also structured around the story of John Cassellis (Robert Forster), the cameraman from the opening sequence, as he documents several unfolding political events for a Chicago television station. While both stories are fictional, the serendipitous precipitation of monumental events during the creation of the film—such as the assassinations of Dr. Martin Luther King Jr. and Robert Kennedy and the riots surrounding the Democratic National Convention in Chicago—became powerful factual anchors. The verisimilitude of these events meant that the fusion of narrative and non-narrative elements within the film was so complete that Wexler built in a series of narrative derailments to constantly remind the audience of the basic fiction behind the potent realism of the sounds and images.

Wexler's working method was to reveal the careful manipulation of "the truth" involved in *cinéma vérité* filmmaking. At every moment when the fiction threatened to collapse into the cultural events surrounding it, Wexler introduced structural "accidents" to remind the viewer that the film was a fictional construction made out of documentary elements. As Stanley Kauffmann noted, "Wexler uses the editing device that Richard Lester used so brilliantly in *Petulia*—cutting suddenly to parallel yet disparate action or to unexplained flashbacks, eventually weaving the pieces together."[33] Examples of this occur throughout the film as characters appear in unlikely locations and impossible situations. In the scene following the title sequence, both Cassellis and his soundman (Peter Bonerz) are seen discussing ethical questions germane to television news reporting with a room full of actual Chicago reporters and journalists. However, in the background at the same event is Eileen (Verna Bloom), a character who appears much later in the narrative and who would seemingly not have any access to such a gathering. A number of these narrative landmines are carefully dispersed through the film to derail the story every time it threatens to be read as documentary realism. As noted by critic Richard Corliss, "The accumulated coincidences and contradictions establish a structure of artifice that almost destroys the impressive realism of the rest of the film."[34] But what Corliss did not understand is that Wexler's entire project was about the creative destruction of this illusion in order to interrupt the story and to force the audience to question the ontological veracity of the *cinéma vérité*-style sounds and images.

In a related strategy, Wexler chose to use only production sound recordings in order to provide a counterpoint to the constantly shifting question of realism in relation to the images. Not only were a system of improvisatory techniques and numerous non-actors used in the film, but the spontaneity of the dialogue and the absolute concordance between sound and image created a contrasting realism to the regular destruction of the film's fictional drive. Wexler hired sound recordist Christopher Newman for his first feature-film recording job after years of working on industrial films and commercials. By utilizing his standard practice of recording the dialogue and sound effects directly, Newman was able to retain all the accidental pauses and gaps in the language that gave the sound track a sense of absolute authenticity: all the sound was recorded live, no dubbing was used, and every line of dialogue was recorded on location.[35] This becomes central to the film as the fictional narrative eventually becomes overwhelmed by the events and documentary realism surrounding the riots at the Democratic National Convention.

When Wexler started writing the film in late 1967, he had planned on making the film strictly about Eileen, a young Appalachian mother, and her son Harald (Harald Blankenship) living in the squalor of Chicago's slums. However, during the film's gestation a number of significant political and social events were worked into a parallel story of the television news reporter. On 30 January 1968, North Vietnamese troops launched the Tet Offensive, a coordinated guerrilla attack on a number of South Vietnamese cities that resulted in massive casualties on both sides. The graphic effect of the "police conflict" was brought home to American television viewers in the horrific newsreel footage of a Vietcong officer being executed by South Vietnamese national police chief Brigadier General Nguyen Ngoc Loan.[36] The widely disseminated Associated Press photograph of the moment of the bullet's impact, coincidentally glimpsed on the wall of John Cassellis's apartment, became one of the most recognizable images of the war in Vietnam and the collapsing control of the American government over the combatants. This sudden lack of confidence in both America's intervention in the Southeast Asian political sphere and the agenda of America's larger foreign policy led to a 31 March broadcast in which President Lyndon Johnson announced that he would not seek or accept his party's nomination for reelection.[37] When the film was completing its pre-production in April, word arrived that Dr. Martin Luther King Jr. had been shot in Memphis. This resulted in Wexler including a scene where Cassellis and his soundman travel to Washington, D.C., to interview several of Dr. King's mourners encamped on the Mall in Resurrection City. And the sudden shock of Robert Kennedy's assassination by Sirhan Sirhan in Los Angeles on 5 June 1968, just days before Wexler started shooting, led to the inclusion of one of the most powerful scenes in the film. In a complete narrative excursus that maintains its chilling effect even after multiple viewings, the film shows workers in a hotel kitchen while Robert Kennedy's voice is heard on the sound track announcing his goal to take his campaign "on to Chicago" and the convention. Immediately after the line is uttered, the fatal gunshots ring out and chaos erupts on the sound track. The audience is kept in a state of confusion as there are neither locational cues to indicate the relationship of the voice to the kitchen nor temporal cues to give evidence to its simultaneity. But seconds after the sound of the assassination and the subsequent aural chaos, the image is disrupted by a re-creation of Kennedy's body being whisked into the kitchen from the auditorium that we realize is just offscreen. The sudden correspondence of sound and image, and the subsequent horror of the event being reconstructed, is immediately broken off when Wexler cuts away at the moment of our recognition, making the momentary correspondence of sound and image even more shocking.

This effect is played out to its fullest extent when the fictional narrative threads come into conflict with the actual riots that arose in response to the 1968 Democratic National Convention in Chicago. Wexler's original draft of the screenplay included scenes of political protests at the convention, but the reality of what happened far outstripped any fictional creation. In a moment of extreme perspicacity, Wexler arranged to film the Illinois National Guard preparing for potential trouble at the convention. The comic counterpoint of reservists playing peace protesters for their uniformed brethren was complexly revisited when Wexler chose to set the film's penultimate scenes against the actual Chicago riots. In a number of graphically vivid scenes where Wexler and his crew follow Verna Bloom as she wanders throughout the barricades and battered protesters, the sequence culminates in the total eruption of the real world into the fictional process. As the crowd begins to retreat and run past the camera, tear-gas canisters release a fog of gas that obscures the lens as an offscreen voice of a crew member is heard exclaiming, "Look out Haskell, it's real!" At this moment, the film creates an absolute concordance between the events unfolding before the camera and the cessation of the narrative line when the exigencies of self-preservation overwhelm the scene's fictional origin.

Despite the film's presumed collapse into documentary realism, the actual story behind the creation of the scene surfaced in 2001 when *Medium Cool* received a celebrated re-release at the Edinburgh Film Festival. In a telling interview with *Sight and Sound* reporter Paul Cronin, Wexler confessed to momentary subterfuge in the creation of the sequence. While it was true that both he and his film crew were tear-gassed at the rally, not every element was in fact a true document of the event. Wexler recalls: "I was out of action for a day and a half. But I have to admit that the line 'Look out Haskell, It's real!' was put in afterwards. It's actually my son speaking the line, recorded months later."[38] In this way, Wexler used the sound track to "lie" to the audience, where the only looped line in the film is the one taken to be most truthful. What becomes so striking is that at the point of absolute cinematic realism within the fictional world of the film, Wexler chose to include this one fabricated line.

By doing so he ensured that the film built in moments of reflexivity even on the level of dialogue to maintain a spectatorial distance from the events and to ensure a dialectical viewing position for the audience. To reiterate this reflexivity, Wexler chose to end the film by adding a further dislocation between sound and visual events. After John and Eileen find each other in the riots, they are seen driving away down an idyllic tree-lined road in a long-take telephoto shot; yet on the sound track we hear a radio broadcast announcing their deaths in a car crash that will occur moments later. After this disruption in audiovisual temporality, the final shot of the film pans away from the flaming wreck, as the sounds of the ongoing riot in Chicago fill the sound track. As the camera continues to pan to the right, it reveals Wexler standing behind a camera in a shot that emulates both the style and ambiguity of the last shot from Godard's *Le Mépris* (1963). According to reviewer Larry Gross: "[Wexler] turns [his] camera toward us as reactions to police violence continue over the credits. This daring self-reflexive sequence is the logical end point of the film's dialectic. Wexler acknowledges his own sense of responsibility for the narrative structure he has fabricated at precisely the moment when real life has brutally gone beyond any rational control."[39] This deliberate breaking of the narrative fourth wall once again reminds the audience that they are being presented with a fictional portrayal cobbled out of documentary elements. And the revelation of the apparatus behind the fictional construction forces them to acknowledge that even those elements that were presented as being "true" need to be evaluated in regard to their operation in the larger narrative framework of the film.

Medium Cool was designed to challenge audiences and to play against numerous spectatorial expectations of a Hollywood fiction film. But another aspect brought the film into conflict with both Paramount and the newly forged ratings system of the Motion Picture Association of America. Although slated for a summer release in 1969, the film came under direct attack from both the studio and the MPAA because of its political content as well as a number of flagrant violations of the then-defunct Production Code. Namely, Wexler had included a scene that featured below-the-waist nudity, the first time it would appear in a major Hollywood release.[40] Because of the visual content and what was deemed to be excessively obscene language, Jack Valenti and the MPAA slapped the film with an X rating.[41] An article in *Time* speculated that the rating was given not only because of the nudity and language, but also in reaction to displeasure on the part of Paramount's owners Gulf + Western, as well as from Mayor Richard Daley's office in Chicago. Wexler replied that he would be happy to replace the obscenities to accommodate his detractors: "I wrote them and said I'd be glad to fix it up. Only I said every time someone said a 'dirty' word I would substitute the word kill. That way we'd have things like 'Kill you!' and 'Put me down, you killer!' I haven't heard any complaints from them since."[42]

Although the film was delayed from release for over half a year, it finally debuted in October 1969, coincident with the trial of the Chicago Seven and the Days of Rage riots sponsored by the Weatherman offshoot of the Students for a Democratic Society. In the heady era of the dissolution of the New Left, Wexler's film resounded as both a declaration of the need for political action and a document of its failure as manifested in Chicago.

For audiences in 1969, who had already witnessed the failure of political protests, the film tapped into a growing sense of discontent that would fully manifest itself in a number of films in the 1970s. *Medium Cool* presented American audiences with the possibility of a politically engaged filmmaker displaying his beliefs directly on the screen. Although a number of youth-oriented films such as *Zabriskie Point* (Michelangelo Antonioni, 1970) and *Getting Straight* (Richard Rush, 1970) attempted to address questions of student protests in the wake of May '68, they did so through either the filter of psychedelic expressiveness or the veil of comedy. Wexler's film not only presented audiences with a complexly structured redeployment of sound and image relations in a blend of fictional narrative and *cinéma vérité* aesthetics, but it was one of the last directly political films to emerge from the faltering studio system.

Puzzle of a Downfall Child (Jerry Schatzberg, 1970)

The history of American cinema tends to overlook the career of Jerry Schatzberg despite the fact that he helmed several small-scale yet bankable films in the eighties (*Honeysuckle Rose* [1980], *No Small Affair* [1984]) and several remarkable films in the seventies. Perhaps this is because filmmaking was not his initial career path. Schatzberg established himself as a leading fashion photographer in the 1960s and was already in his forties when he decided to start directing commercials before working as a motion picture director.[43] Although Schatzberg was very much an outsider to Hollywood, and not even aligned with the rising New York–based directors of the day, his connection to the new American cinema comes directly from *Bonnie and Clyde*: it was the influence of his fiancée Faye Dunaway who convinced him to try his talents at motion pictures after her breakthrough success in Penn's film. For his first foray into moving images, Schatzberg decided to stick close to what he knew by crafting a story of the rise and fall of a fashion model. The storyline and dialogue of *Puzzle of a Downfall Child* were based on recorded interviews that Schatzberg made with a close friend, the fashion model Anne Saint Marie, about her mental breakdown and whose experience became the narrative kernel for the film's main character, Lou Andreas Sand (Faye Dunaway).[44]

Puzzle of a Downfall Child represents a curious blend of cinematic styles. On the one hand, its high-key lighting and gauzy visual style borrows from fashion photography. On the other hand, the influence of the British New Wave directors, especially Lester and Schlesinger, can be traced through the sound and editing techniques employed by Schatzberg in the film. *Puzzle of a Downfall Child* bears testament to the influence of the sound experiments from British New Wave as well as the liberation and freedom gained from shooting on location witnessed in *Bonnie and Clyde* and *Medium Cool*. The film's screenplay was crafted by Carole Eastman (under her *nom de plume* Adrian Joyce)[45] and borrows heavily from Schatzberg's interviews with Anne Saint Marie. The resulting film is a jarringly powerful look behind the scenes into the world of high-fashion photography and the corrupting power of success.

It is also an important hybrid of British and American sound techniques that points to developments later in the 1970s and serves as a template for films that foreground the recording of the female voice.[46] The parallels with the British New Wave directors are evident right from the start, where the first lines of "Why don't we begin . . ." and "Is this being recorded?" over a shot of a desolate beach house in winter make the viewer aware of the forthcoming dissociation of sound and image. The story begins with Lou being interviewed by her long-time friend and former lover Aaron Reinhardt (Barry Primus), a fashion photographer. As the camera progressively cuts to closer views of the beach house in a clear homage to *Citizen Kane* (Orson Welles, 1941), right down to the "No Trespassing" sign, the voices carry on at the same level as the sequence ends on shots of the VU meter and turning reels of tape. Over the images we hear Lou's voice saying, "I'll tell you something you don't know," where she reveals that her real name is not Lou Sand but Emily Versene. Right from the beginning the issues of fact and fantasy, documentary and fiction, are foregrounded and the issue of truth is framed as a subjective variable.

Aaron's interview provides the springboard for numerous flashbacks to Lou's past that start to describe her rise and decline as a model. This is the same narrative conceit that Schlesinger used in *Darling* (1965), yet instead of the conflict between the recorded description and the actual events that occur in the flashback, here the recordings become a transcript of Lou gradually losing touch with reality. As Lou continues to describe her life story to Aaron, there is a progressive separation or disjunction between the voiceover and the images that shows her mental decline over the years. Moreover, the presence of the recording device sets up a narrative expectation that throughout the course of the interview more of the truth will be revealed. Ironically, what is revealed is Lou's growing inability to function on a day-to-day basis, and the film introduces a main trope that dominates American cinema (and culture) in the 1970s: the ubiquity of recording devices and their effect on their subjects.[47]

The tape recorder functions as the central narrative device in the film—allowing for the triggering of flashbacks and excursions into Lou's memories—and it serves as an aide-memoire in the evaluation of the testimonies it records. One of the main acoustic effects heard during the recording sessions is the attention given to the exterior ambiences. As Lou and Aaron discuss the past, the external ambience changes to signify shifting attention levels. Similar to *This Sporting Life* (Lindsay Anderson, 1963), the levels of ambience generally fade out after the flashback begins and its end is signaled by the encroachment of the beach ambience and voiceover. In addition, the flashbacks themselves contain nested flashbacks that stem from Lou's unreliable recollections of the past. As *Puzzle of a Downfall Child* progresses, the boundaries between flashbacks, flashbacks within flashbacks, misremembrances, and fantasies break down. Initially Lou's voiceover does not continue in the flashbacks, yet once the flashbacks within a flashback are introduced through her voiceover it allows the film to jumble the present narration with imbricated flashbacks and delusions. The pattern of sound bridges and advances sets up a strong acoustic structure in the film that guides the spectators into and out of the flashbacks, yet it does not let us distinguish between the real remembrances and the fantasies.

Indeed, as indicated by the title, the structural ordering of the film resembles a puzzle and our job as viewers is to reassemble the details of Lou's life into a coherent order while accounting for the elements that don't fit into the logical progression of the story. These dissociative elements—the byproduct of Lou's imagination and growing delusions—are markers of a blurring of reality and a dissociation between sound and image closely linked to both *Petulia* and *Midnight Cowboy*. Approximately one-third of the way through the film Lou has a moment of complete dissociation when her conversation with Aaron is intercut with images of her walking along the beach accompanied by the mute fisherman, Mr. Wong. As she continues to stare out the window of the beach house, talking to Aaron, we see her walking on the beach saying the very same words to Mr. Wong. What is immediately striking is that, although she is seen in extreme long shot with the surf pounding behind her and her words are synced to her one-sided "dialogue" with Mr. Wong, Lou's voice continues to be heard in close-up as if spoken to Aaron in the cottage.[48] According to the codes already established in the film, it is fair to assume that most audience members read this as another flashback, yet the radical difference in sound and image scales signals that something else is happening. For in the last shot in the sequence Lou walks away from the fisherman as the camera tracks back to reveal her looking out the window at herself on the beach. This obviously impossible shot construction signals that some, if not many, of the stories and remembrances Lou has conveyed may not actually be true.

The stabilizing element in the film and in the soundscape is Aaron's voice, which constantly prompts Lou to return to the present and to stay on the side of the factual. While this pattern is established early on, it allows for interesting variations in sound and image relations when Lou does not want to reveal elements of her past. Specifically, Lou dodges a question about a time when she drifted out of Aaron's life by coquettishly playing the castanets and rambling on about the source of her atheism. When Aaron presses her again to recount the period, Lou requests that he play back the tape so she can hear herself play the castanets. As the recorder rewinds and replays the prior two minutes of conversation, the image displays Lou having a tryst with a man she picked up in a bar. The images cut abruptly from the bar to Lou and the man making love in a hotel room to Lou staring at her reflection in a grossly distorted mirror. The attenuated sound of Lou's and Aaron's voices on the tape recording are the main sounds heard on the sound track, accompanied by Michael Small's minimal electronic score, as the images of the affair transition into a much younger Lou praying in a ruined church. Only two sounds filter through from the flashback to join with the sounds of the tape recording: a moment when Lou leans over to the man at the bar and says, "Emily," her birth name, and the sound of the door closing when her lover leaves the next morning. This condensation of sound and image is linked to Lou's replayed dialogue where she explains, "I now know I was always atheist because of the idea of punishment for being what you are or what you may think." This dialogue, which initially was heard as Lou being evasive, now takes on a new meaning when set in counterpoint with her loss of religion as a child and her inability to love. The sequence ends abruptly as Lou is summoned back from her mental images by Aaron's voice on the recorder and the shot of the church is peeled away to reveal an emotionally jarred and shaking Lou.[49]

The divergence between Lou's conversation with Aaron and her memories increases throughout the film, blending elements of fact and fantasy into an increasingly fragmented mix. As Lou describes a dinner party held by her socialite friend Pauline (Viveca Lindfors), the camera displays a large table set for a dozen guests with only Lou, Pauline, her husband Uwe (Barry Morse), and their friend Mark (Roy Scheider) present. In contrast, the "walla" sounds of a much larger party are heard on the sound track,[50] making the audience aware of how Lou's memory is filtering the events of the evening. If that were not clear enough, when Lou explains in voiceover, "[Pauline] would act so insimulated that her guests would get up and leave," the walla track cuts out entirely, leaving just the sounds of the four guests in an empty room. Later, when Lou and Aaron finally do have a brief romantic interlude, Lou insists on role-playing and instructs Aaron to pretend they don't know each other when they arrive at a bar. Intercut with Lou's instructions are various versions of the encounter.

At first Aaron cannot step into his role and asks Lou at the bar, "What am I supposed to say?" only to be "answered" by a cut back to Lou's instructions in the car. This pattern carries on until Aaron asks Lou to go with him to a motel. His fumbling appeals and shaggy-dog story about his "two-ton yellow truck" are heard on the sound track over images of them arriving in the motel room. The scene mirrors Lou's earlier pickup in a bar, and at the moment Lou and Aaron start making love the same electronic score returns with a brief flash frame of a figure from Lou's past:

> For a brief moment we see an image of an older man who, we learned, raped Emily when she was a teenager. This is effected in a subjective point-of-view shot, from an extremely low angle, as the man lowers his face toward the camera. The shots are cut together in a shot/reverse shot sequence to convey the idea that the adult Lou suddenly becomes her younger self and sees Aaron as both a scripted "stranger" and the older man she once knew. The implications are complex and ambiguous. . . . Lou's fantasy now seems to function as a form of repetition-compulsion.[51]

Instead of another flashback to her youth the audience is forced to watch Lou's pained expressions and ineffectual utterances as Aaron ignores her protestations. The fragmented editing patterns, the repeated scenes, and the fantasies signal Lou's deteriorating mental state to the audience.

Although the film plays with reflexivity, it does so within a system of its own design. The contrasts between Lou's visual memories and her spoken reminiscences are not designed to distance the spectator from the story. Rather, they are read as markers initially of Lou's inability to distinguish between fact and fantasy and eventually as her downfall into mental illness. Later scenes confirm this when Lou shows Aaron her white wedding dress at the beach house, yet her flashbacks to her failed relationship with Mark show her wearing a black dress when she runs out on their wedding. When Lou's delusions begin to overtake her she attempts to seek solace in the arms of Pauline's husband, Uwe. As he resists her advances their dialogue continues over an impossible flashback of a young Aaron witnessing the scene from Lou's past when she first encountered the man who raped her. In a shot/reverse shot pattern that inverts the pattern established in the motel, Lou looks back at Aaron to see that he has disappeared from her memory when Uwe says, "You're ill, you understand?" As the sounds of Uwe's apartment fade out, the image cuts to a medium close-up of Lou lying in a field followed by the same flash-frame shot of the older man seen from below. As she recoils, a scream is heard and slow motion images of Lou being led, straitjacketed, into an asylum are accompanied by reverberant pants and moans as she struggles in vain. As suddenly as the sounds began they cut out to reveal Lou in a white robe, sitting in a white hospital room, asking Uwe not to tell anyone that she's been committed. The conversation continues, in shot/reverse shot, until Uwe is unexpectedly replaced with an anonymous orderly who carries on the conversation. Lou's fragmented grasp on reality is displayed as her conversation fluidly transitions from one imagined character to another—first Uwe, then her psychologist, and finally Aaron—interspersed with flash cuts of her actual agitated state.

At the end the film returns to the present after Lou, delusional in her hospital room, explains to the imagined Uwe that "everything was in my mind, that's all!" Over an image of her in the hospital, drugged and literally out of focus, we realize that the dialogue has actually transitioned back to her conversation with Aaron at the beach house, telling him that "now things will be much different—I'm trying to be very good." The final scene as Aaron prepares to leave restores a sense of stability, of both Lou's character and her story. In the penultimate shot there is a direct concordance between sound and image temporalities during a tour de force four-minute take as Lou and Aaron walk along the beach to the ferry. In order to film the characters, Schatzberg and director of photography Adam Holender used a long telephoto lens and lavalier microphones with radio packs to record the dialogue. The resulting shot is in striking dissonance to the tight framing and boom mic recordings used throughout the rest of the film. As Lou and Aaron walk along the beach, braced against a steady breeze, their voices are heard directly, without distortion from the wind and perfectly in sync. When Lou asks why she and Aaron never had an affair Aaron reminds her that they did, and Lou breaks down in tears, happy when she realizes that what she thought was fantasy was in fact true. The verisimilitude of the dialogue and the character movements matches the narrative return to normality and Lou's newfound lucidity. Despite the synchronization of the voices, however, the discord between the shot scale of the telephoto lens and the close-up sound scale from the lavalier mics creates a disturbing undercurrent of doubt. Such a dramatic rupture of sound and image scales occurred only once before in the film, during Lou's fantasy conversation with the fisherman on the beach, and the sound of the shot hints at the very likely possibility that Lou will slip back into delusions as soon as Aaron leaves. In confirmation of this the plane of focus stays in the foreground during the last shot when Aaron leaves Lou behind. As she returns to the cottage and recedes into the background of the frame, Lou grows more and more out of focus, and to reinforce the effect the frame matte grows smaller around her while the end credits scroll over the image.

The sound and image aesthetics in *Puzzle of a Downfall Child* exemplify a unique fusion of a willingness to experiment with sound and image relations imported by the British New Wave directors combined with an empirical probing of the codes of realism drawn from documentaries and television. The result balances a need for realism with an ability to accommodate character subjectivity and reflect emotional changes through sound. Schatzberg continued to refine his personal style throughout the 1970s, and the next two films he made—*Panic in Needle Park* (1972) and *Scarecrow* (1973)—reinforced a strong sense of acoustic realism through the exclusive use of location sound recording and a refusal to include nondiegetic score music. Yet none of his films would be as boldly structural in their approach to sound and image relations as *Puzzle of a Downfall Child*, and few directors were able to craft personal sound aesthetics that utilized the sound technology in radically new ways. Of those who did in the early part of the decade, Robert Altman, Alan J. Pakula, and Monte Hellman stand out as exemplars of emerging personal sound aesthetics in support of narrative storytelling.

4

New Voices and Personal Sound Aesthetics, 1970–1971

At the vanguard of new sound practitioners were a group of experienced filmmakers making their way into major studio filmmaking. Building on the sound techniques pioneered in the 1960s, directors such as Robert Altman, Alan J. Pakula, Monte Hellman, and George Lucas each developed personal sound aesthetics based around his experiences working outside of or tangential to the Hollywood system. Altman, after an early foray into feature filmmaking, spent his formative years working in and out of Hollywood and television for most of the 1960s. Pakula started producing films for Robert Mulligan in the late 1950s, including *To Kill a Mockingbird* (1962), before branching out as a director in 1969 with *The Sterile Cuckoo*. Hellman cut his teeth working for Roger Corman and directed a string of low-budget independent films in the 1960s before he made *Two-Lane Blacktop* for Universal in 1971. And Lucas began as a film student at the University of Southern California where he created several experimental 16 mm films; his short *Electronic Labyrinth: THX 1138 4EB* won the top prize at the 1967–68 National Student Film Festival. Even though each director came from a different background, they each cultivated sound strategies in their films that marked a major change within the Hollywood system. After the failures of several large-budget films and the runaway success of films like *Easy Rider* (Dennis Hopper, 1969), the studios were willing to take chances on unknown and untested directors by green-lighting lower-budget projects. The result was that within the studio system directors were experimenting with new representational models and discovering new approaches to sound and image interaction.

*M*A*S*H* (Robert Altman, 1970)

One of the most important formal experiments in the films of Robert Altman was multiplying the number of central characters to hyperbolic proportions while simultaneously ensuring their ability to be seen and heard through a sophisticated blend of zoom cinematography and radio microphones. After a brief stint working on industrial films and documentaries in the 1950s, Altman embarked on a successful career as a television director. Between 1957 and 1966, he directed episodes for shows as varied as *Alfred Hitchcock Presents, Bonanza, Gunsmoke, Combat, Kraft Mystery Theater, The Whirlybirds, Peter Gunn*, and *Route 66*. Although his experience in television was rather scattered, having worked on nearly two dozen different shows, yet never for very long, Altman developed a keen sense for dramatic timing and the value of improvisation. Crossing back to feature filmmaking with the film *Countdown* in 1968, Altman borrowed a few techniques from his work in television to accomplish his goals. First, he would shoot with a camera using a zoom lens to allow for the flexibility of changing shot size in mid-take to accommodate his large, ensemble casts. Second, he valued the use of overlapping dialogue to include a plethora of narrative information on the sound track. Evidence of this technique can be heard emerging in *Countdown* and the conflict between the central narrative line and a series of "subconversations" in *That Cold Day in the Park* (1969). But Altman's cinematic techniques came to fruition with the January 1970 release of *M*A*S*H*.

Luck clearly had more to do with Altman's getting the job than any cinematic acumen or clout; rumors place him as the fifteenth person asked to direct the film after fourteen other directors declined.[1] While it was hardly a glowing endorsement, Altman took the job and proceeded to dispense with Ring Lardner Jr.'s screenplay in favor of a series of improvised vignettes and a narrative framework that tore apart the linear three-act structure of the original script. The combination of two seemingly unrelated cinematic approaches is what made *M*A*S*H* such a strikingly different film for audiences upon its initial release. Not only was Altman able to cover entire scenes in a single take with his wandering zoom lens, but the use of radio microphones and careful live mixing, orchestrated by the director, also meant that even the most banal scenes became complex audiovisual symphonies. Instead of using the zoom as a standard substitute for a forward or backward tracking shot, Altman used it as a tool to accommodate the improvisation of his actors and to bring the presence of time back into the shot. According to film scholar Robert Kolker: "The visual structure of [Altman's] films requires not that the viewer pick and choose among various visual and aural options but that he or she observe and understand the whole and integrate into the larger unit those parts of the whole that the director wishes to emphasize. What Altman creates is not the conventional structure of a whole that is analyzed into its parts, but a simultaneity of the whole *and* its parts, a simultaneity the viewer must always attend to."[2] As Kolker pointed out, the notion of simultaneity is critical in Altman as it represents an emphasis on the plurality of events occurring on the screen rather than a single central narrative event or character. What results is that the audience is placed in an active spectatorial position, being asked to sift through the audiovisual information in order to follow one of many potential narrative paths. While this is similar to the ambiguity that André Bazin noted in the deep focus constructions of Orson Welles and William Wyler,[3] Robert Altman expanded the freedom of the spectator by including a number of competing conversations in each scene. This rejection of the standard narrative convention of one character speaking at a time in favor of multiple, overlapping voices had the unlikely effect of bringing a new sense of realism into the scenes. As Altman himself explained, "That's to give the audience a sense of the dialogue, the emotional feeling, rather than the literal word. That's the way sound is in real life."[4]

Altman's unique aesthetic approach to *M*A*S*H* was rooted in his emphasis on the improvised performances of the actors involved in the production. By sequestering the cast and crew on location at the Fox ranch in Malibu Canyon, Altman was able to create an environment whereby all the actors were asked to contribute new elements to scenes. But unlike the intricate rehearsals used by filmmakers like John Cassavetes, or the addition of subconversations and dialogue during post-production in the films of Richard Lester, Altman preferred to use minimal rehearsals and to allow for the possibility of live "accidents" and spontaneous reactions. David Denby explained the basic approach used on *M*A*S*H*:

> Rather than scheduling his cast members to fly in for three days here, four days there, as is usually done, Altman gathered an ensemble company, mostly at low pay, and held his cast together for the entire six weeks of shooting. Many of the actors were hired from the American Conservatory Theater and the satirical revue *The Committee*, two San Francisco institutions whose members specialize in ensemble playing and improvisation.[5] Eight of them had no parts in the script when hired, but Altman put them in costume and let them join the group on set. As tensions and friendships developed, roles emerged for these actors and lines and routines for the company at large—just as they might at a close, working community like an Army base. Some of the actors developed such a strong sense of the community that they voluntarily removed themselves from scenes where they felt their characters didn't belong.[6]

To facilitate this communal effect in the film, most actors were wired with hidden radio transmitters and lavalier microphones to allow for their free movement on the set and to eliminate potentially obtrusive cables and booms. Altman was one of the first directors to use the newly transistorized Vega radio microphone units, which appeared on the market in 1969. Before the introduction of the pocket-sized radio units, most radio packs were bulky with very limited ranges. Moreover, they were extremely susceptible to interference and breakdown. The choice of Bernard Freericks as sound recordist was especially significant due to his longstanding relationship with radio microphone use.[7] As each take unfolded, Freericks would mix the multiple conversations live, sending a feed to the director. While this freed the characters to move anywhere and to carry on multiple conversations, it also meant that there was a need for the camera to follow the characters around the set. Altman kept two cameras running during each scene "to keep the actors honest" since they never knew whether they were being filmed or not.[8] Cinematographer Harold E. Stine heightened the effect by shooting each scene with a constantly moving zoom lens. As characters moved about the scene, or as conversations overlapped, the cameras would adjust to follow the shifting narrative focus. Behind all this was Altman "conducting" the cameras, instructing Stine and his camera operators to follow specific characters as he listened to the live audio mix on headphones. Even though this often resulted in a high shooting ratio, the speed of the production process and the lack of retakes due to the comfort of the actors meant that the film actually came in under budget.

As a result, the film stands in striking contrast to other films from 1970. The use of the zoom lens was an expedient way of shooting entire scenes in a single take, but it also created a unique effect within the film. According to Robin Wood, "For Altman, the zoom is at once his means of guiding the audience's consciousness and of asserting his own presence in the film; but he has also grasped its potential for dissolving space and undermining our sense of physical reality."[9] To address Wood's first point, the presence of a constantly moving zoom frame is radically different from the standard fixed-camera constructions of Hollywood filmmaking and is sufficiently different from the "invisible" tracking shots used to great effect in the films of Jean Renoir and Alfred Hitchcock. The tracking shot has been identified as a method for revealing space outside the frame in a way that creates continuity throughout the shot, yet Altman's zooms function in a very different way. Whereas Hitchcock's tracking shots were used to create a conceptual linkage between the rigorous patterns of viewing in his films and significant narrative information, Altman's zooms constantly remind the audience of the shifting focus of attention in the film and the presence of the film's author, who is subtly guiding the audience along. As John Belton noted, "Altman uses the zoom to assert his own narrative voice, frequently relying on it for transitions. Altman's zooms function like jazz improvisations superimposed over a fixed melody: whether motivated or not, they signal his presence as a narrator."[10] Although the scenes are always open to the contingency of interpretation, Altman regularly limited potential overinterpretations with carefully placed insert shots to reposition the spectator's attention, thereby creating narrative pathways both external to the original shooting script[11] and as a byproduct of the editing process.

Regarding Wood's second point about the zoom lens "dissolving physical reality," in Altman's films this becomes more a function of the sound. Similar to the way that the use of the mobile zoom lens and widescreen frame offers a shifting form of visual presentation, the effect of the closely miked radio microphones creates an "acoustic foreshortening" that emulates the flatness of the telephoto image. Because the use of lavalier microphones in conjunction with the radio packs effectively eliminates the depth cues that normally come from reverberation, the spectator/auditor is forced to hear all the voices weighted equally in a scene.[12] The absence of acoustic depth cues in Altman's films means that the audience has to listen for differences in volume or shifts in the relationship between the voices and their position within the frame. Instead of hearing the main conversation miked directly, usually from a boom just above the actors, and the background conversations as an indistinguishable field of walla, Altman regularly gave the audience two or more competing conversations in a given scene. As an example, the scene of Duke (Tom Skerritt) and Hawkeye's (Donald Sutherland) arrival at the camp contains as many as three conversations occurring at the same time. While this does create a sense of confusion in the inattentive spectator, careful analysis reveals that as the camera pans and zooms to emphasize the different characters, the relevant subconversations become legible through the audience's ability to hear the words while seeing the characters speak. A zoom in and pan right forces Hawkeye and Lieutenant "Dish" (Jo Ann Pflug) out of the frame, and although the conversation between them continues while offscreen, the camera reveals Lt. Col. Henry Blake (Roger Bowen), whom we hear asking whether Hawkeye is the new surgeon they've been expecting. While Altman could have handled the scene in two sets of shot/reverse shots, the integration of both narrative elements into the same mobile frame creates a much stronger sense of unity and enforces the effect of audience participation by not fragmenting the profilmic space.

This approach to filming allows for an emphasis on the formal unity of a particular scene and its relationship to the narrative. A significant effect of this is the deemphasis of the normal codes of mise-en-scène, which dictate that the characters should always be presented directly to the camera. Instead, Altman regularly has the camera shooting through the gauze of the tents, around barbed wire and debris,[13] and through dust and smoke so the spectator has to "search" for the narratively significant details within a scene. To augment this effect, Altman also keeps the sound track consistently busy. As Kolker noted,

FIGURE 6. One of the many ubiquitous "squawk box" loudspeakers seen and heard throughout *M*A*S*H* (Robert Altman, 1970).

"There is not a silent moment in *M*A*S*H*: dialogue, music, announcements on a loudspeaker are continuous, sometimes at odds with, or in ironic counterpoint to, what is happening on the screen, sometimes all things at once."[14] Often Altman used this disjunction to subvert the expectations of the audience in order to create comic juxtapositions between the sound and image tracks—the broadcast of Hot Lips (Sally Kellerman) and Frank's (Robert Duvall) lovemaking session to the camp, Radar's (Gary Burghoff) "parroting" of Colonel Blake[15]—but Altman's working methods were significantly more complicated than just the dislocation of sound and image for the sake of a gag.

As a final way of disrupting the audiovisual order of the film, Altman built in one continuous sound motif that undergirds the disjunction between sound and image while also literally manifesting the authorial voice within the film. In addition to the constantly overlapping dialogue, announcements and music are regularly heard through the "squawk box" speakers positioned throughout the camp (see figure 6). Even though none of the character voices have any sound scale in relation to their diegetic position, the announcements and music are spatialized due to the addition of reverberation. This ontologically realistic use of the speakers is undercut when, for the final images of the film, the speaker recites the end credits aloud as the images unfurl on screen. In the words of Robert Altman:

> The picture is satire on satire. *Time* magazine had a great review . . . that caught what we did with the end of the picture, that in order to make the real, full statement, we had to say this picture is also full of [shit]. Anyway, *Time*'s review really caught the point that pleased me the most . . . that the Squawk Box becomes as important as any character in the film, and that after destroying everything in sight, it turns on itself and takes

its own life, and to me that's it. I don't think the film could be the same without it, if it ended in any other way.[16]

Here, the one secure position of realistic audition in *M*A*S*H* is deliberately undermined to emphasize the complete manipulation of the codes of representation by the director.

Altman's methods continued to take shape during the course of the 1970s as all his films attempted to redefine a relationship between the audience and genre by subverting the audiences' expectations of stable generic forms. This systematic approach to genre reinscription can be seen across a series of films including *McCabe & Mrs. Miller* (the western), *The Long Goodbye* (the detective film), *Thieves Like Us* ("lovers on the run"), *California Split* (the buddy film), and *Nashville* (musical; political film). By treating genre conventions as being as malleable as narrative conventions, Altman displayed the fact that all systems of cinematic representation were entirely contingent on audience conditioning and expectations. The larger project presented in the scope of Altman's films is one of rewriting the rules of cinematic spectatorship. As John Belton wrote about Altman's use of the zoom lens: "[Both] spatially distorting and inherently self-conscious, the zoom reflects the disintegration of cinematic codes developed before the Second World War. . . . Space is no longer defined in terms of perspective cues and parallax, but in terms of changing image size and time."[17]

To Belton's argument it is possible to add the fact that the use of lavalier microphones also disintegrated the acoustic cues of reverberation and volume changes, leaving space to be determined purely in the *opposition* of sound and image tracks. If there was one problem involved in Altman's early system of microphone use, it was the inability to mix dialogue when it was recorded live during a master take. Even this problem was solved when sound recordist Jim Webb introduced the use of the portable Stevens eight-track recorder during the filming of *California Split* (1974), thereby allowing Altman control over all the individual dialogue tracks in post-production. This "unmixed" sound made possible the truly democratic voice as evidenced in *Nashville* (1975).

Klute (Alan J. Pakula, 1971)

Alan J. Pakula distinguished himself from his peers in the 1970s by the odd fact that he moved from being a producer to direct three of the decade's most memorable films: *Klute*, *The Parallax View* (1974), and *All the President's Men* (1976). Often called his "paranoia trilogy," these three films explore the darkest corners of American life during the decade and alternately find despair and hope in equal measure. Of the three, *Klute* stands out because of its unprecedented style and sophistication, not to mention a central narrative construction that would shape all three of his paranoia films: surveillance. A full year before the Watergate break-in shook the foundations of American democracy, Pakula's intimate tale featured the use of surreptitious recording as the linchpin of its narrative. The film follows John Klute (Donald Sutherland), a small-town private detective, who has been hired to follow up some leads about his missing friend Tom Gruneman. The leads take him to Bree Daniels (Jane Fonda), a part-time model/actress and full-time prostitute who, despite the misleading title, figures as the story's center. What unravels is not just a simple "whodunit" because the solution to Gruneman's disappearance is revealed early in the film and a careful viewer can extrapolate the answer from the very first shot of the film. However, the way that the film is constructed and the complex presentational strategies mapped out in both image and sound use make this one of the most intriguing films of the decade.

Released on 25 June 1971, *Klute* was Pakula's second film as a director but it showed a remarkable sophistication and style and an awareness of how the codes of genre could be altered to augment the story. During a decade and a half working as producer for Robert Mulligan, Pakula learned the mechanics of filmmaking, an education that naturally enough led him to try making films on his own. After a brief misstep two years prior with the uneven *The Sterile Cuckoo*, Pakula found his own directorial voice in the taut thriller that was *Klute*. The missing person investigation, routine in its structure, was turned by Pakula into a mesmerizing essay on modern morality and the loss of privacy in a technocratic society. Central to this is the use of tape recording and surveillance as the principal narrative device of the film. Specifically, the film's villain records his crimes on a portable tape recorder and revisits these recordings throughout the film. As a result, the spectator is forced into a new relationship between the sound and image tracks, where sound is regularly used in counterpoint to the images or to introduce a sense of ambiguity into the story.

To achieve this overall strategy, Pakula developed an audiovisual system of representation where nearly every scene is introduced through sounds first while semi-distinguishable images are seen on the screen. On a more formal level, the director uses this strategy to create an effect based around the regular "acousmatization" of the voices. According to film theorist Michel Chion, the "acousmatic voice" is one that hovers just beyond the frame of the image, threatening to enter the scene, but never being fully part of either the diegesis or the realm of narratorial voiceover. "A sound or voice that remains acousmatic creates a mystery of the nature of its source, its properties and its powers, given that casual listening cannot supply complete information about the sound's nature and the events taking place."[18] The effect is one where the spectator is uncertain of the ontological relationship between the sound and the image, and in that gap enters a disturbing suspense. This momentary suspension of synchrony between sound and image subsequently creates a sense of

FIGURE 7. The theme of surveillance is introduced at the beginning of *Klute* (Alan J. Pakula, 1971) when a small tape recorder is glimpsed briefly on the Grunemans' Thanksgiving dinner table.

uneasiness in the audience where the nature and source of the sound is constantly in question. This tactic is a powerful device used by Pakula throughout the film and it is inscribed as a pattern from the very first frames.

In the opening shot, the viewer sees a small running tape recorder sitting on a table, and to its side an out-of-focus fruit bowl with bananas and a glass of wine can be made out (see figure 7). The presence of the tape recorder is unexplained, and as subsequent cuts reveal the scene to be a gathering of friends at what we discover to be the Gruneman household, we are kept in suspense due to the lack of any locatable synchrony between sound and image. It is only after the camera tracks to the far end of the table that we hear Mrs. Gruneman's voice and realize that it is in sync with her lip movements. Up until that point the presence of the tape recorder could indicate that the sounds we are hearing are synchronous with the events and being recorded, or, as in *Puzzle of a Downfall Child* (Jerry Schatzberg, 1970), its presence could trigger the images as a flashback when it replays a prior event. Yet as soon as the audience determines this synchrony, it is broken. We hear a toast to the Grunemans as Mr. and Mrs. Gruneman raise their glasses to each other in shot/reverse shots, but the film then breaks the pattern, cutting back to Mr. Gruneman's empty chair seen at some later time. We suspect that the absence of the character is narratively motivated, but we discover this only when we hear an unfamiliar offscreen voice saying, "Did you know the subject Thomas Gruneman?" Slowly, over the next few shots, we are able to discern when this is, who is talking, and the fact that Mr. Gruneman has been missing for a period of time. In this scene and throughout the film, cinematographer Gordon Willis placed his camera so that the characters are either too close to the lens to stay in focus, partially outside of the frame line, or with their backs turned to the camera to obscure their faces when they talk. By keeping the figures of the FBI officers hidden through editing and visual masking, Pakula enhances the mystery of the acousmatic voice.

Though this procedure does provide a narrative function, adding suspense to a scene by not allowing the audience to know who is speaking, it also creates a larger structural effect in the film, one based around the symbolic power of the acousmatic voice. Specifically, these elements are constituent to the creation of the paranoid subject in a system of postmodern cinematic representation. As mentioned by Fredric Jameson in his exhaustive analysis of a particular trend in the conspiratorial narratives of 1970s cinema, the function of representing social totality regularly gets mapped onto the conflict between the public and the private spheres. In the case of *Klute*, this is manifested in the theme of the epistemological act of detection and its opposite in the ontology of conspiratorial totality.[19] Part of the nature of the traditional structure of ratiocination is that it always presupposes a subject—the murder—and a telos—the reconstruction of the events leading up to the murder. Although both elements are present in *Klute*, the film is far less about the act of detection in the traditional sense than it is an essay on the nature of human existence in the contemporary city. According to Jameson, the film's sound strategies are in keeping with a shift away from the representation of visual reality. It is best taken as a "moment in the historical process of postmodernization, in which the decisive modulation from the visual image to the auditory one is as fundamental as it is paradoxical."[20] This move to the acoustic register in the postmodern allows for the center of the conspiracy, the object of the panoptic gaze, to be rendered through the voice. Moving from the specular to the auditory, *Klute*'s narrative structure echoes this logic in the way that the control over the human voice is configured as the locus of power in the film.

This is demonstrated through the narrativizing of the recording apparatus and its effect on the diegetic listening subject. Whereas *M*A*S*H* created a partial awareness of sound technology through the use of the loudspeakers as a Greek chorus, providing strophe and antistrophe to punctuate the central actions, *Klute* entirely foregrounds the procedure by manifesting the recording apparatus in the diegesis. Often the recording itself takes precedence over the images, especially in the enigmatic title sequence. Here, the same tape recorder seen in the opening sequence reappears, resting against a completely black backdrop while a pair of hands starts its playback. The entire title sequence consists of the tape replaying an unknown female voice, whom we soon realize is that of a prostitute relaying the technical details of an encounter with a john. What is unnerving to the audience is the fact that the other voice in the conversation, that of the john, is conspicuously absent—presumably edited out to preserve his anonymity. This enacts a double process of acousmatization in the sequence. First, we become aware of a female voice that we have not previously heard and desire to reveal the body that is speaking. But because this is being presented as a recording, a definite notion of time is also injected into the scene. We are encountering a prior event, one that has taken place before this replaying of the tape, which is also apparent because the body of the acousmatic speaker is entirely absent from the scene. However, it is the second, unheard voice that marks the double acousmatization of the tape as we are aware of the man's presence in the diegetic space, yet his voice is neither heard on the tape nor in the room. The narrative reasoning for this is that the playing of the man's voice on the tape, for the attentive listener, would give away the identity of the murderer. However, in the larger function of the film, the absence of the man's voice marks the control over his identity by the editing of the tape, and a larger measure of control over the mysterious speaking voice on the recording.

Of course, to most filmgoers in 1971, the mystery voice would have easily been identified as that of Jane Fonda, a fact that is corroborated in the next scene when she is seen, but not heard, at a modeling audition. Pakula deliberately keeps her physical presence in the diegesis of the film and her voice separated for several minutes of screen time. After the title sequence, the film cuts back to the Gruneman house where his boss, Peter Cable (Charles Cioffi), and a family friend, John Klute, are being briefed by two FBI agents about their failure to turn up any clues about Tom Gruneman's disappearance. As one of the officers explains that they had kept a prostitute, Bree Daniels, under surveillance after she received some obscene letters from Tom Gruneman, we see her at the audition. As the agent details the brutal facts of daily life for a call girl, we see Bree walking down a New York City street and stopping at a phone booth. This sound bridge and overlap lasts for thirty-five seconds of screen time, emphasizing the conflict between the visual and acoustic tracks, but also introducing Bree by means of the controlling voice of the officer. This sound overlap motif is structured throughout the film to indicate control of the speaking voice over the image appearing "behind" it.

Power is consistently associated with the *acousmaître*, Chion's portmanteau for the master of the acousmatic voice, and the ability to speak while unseen.[21] Pakula "acousmatizes" characters by giving them the privileged speaking position of either the sound overlap or the sound advance at the start of scenes. But it is the central function of the tape recorder that grants certain characters the ability to capture and control the voice. In this way, the tape recorder operates as what Jameson refers to as an analagon, "a quasi-material object of perception off which we read, as from a material *interpretant*, the narrative language of another set of events."[22] By using the recording apparatus to convey the horror of the attacks, Pakula never actually has to render them visually. It is all the more terrifying because of the complex system of analagons that Pakula places throughout the film. Telephones, psychiatric sessions, police interrogations, and theatrical auditions all function to remind the viewer how the voice is used to control and condition Bree as well as other women in the film.

To create a visual analog for the unrenderable quality of human speech, the film replaces the absent male speaker with the image of the tape recorder. This shifts the emphasis to the means of reproduction, literally foregrounding the apparatus and making the audience aware of the manipulation involved in the recording process, both within the story itself and in the cinematic act of production. Importantly, the drive to see who is speaking and what is actually happening provides a secondary operation within the film that eventually supplants the narrative cause-and-effect chain. By systematically denying the spectator the "unveiling" of the acousmatic speakers, Pakula introduces instability in the text that denies the closure effect and keeps propelling the film to its conclusion.

The mystery surrounding Gruneman's disappearance is solved a quarter of the way through the film when it is revealed that the stalker is actually Peter Cable. Two early shots hint at this when he is nearly caught by Klute spying on Bree's apartment and he is later seen observing her from across the street. But confirmation occurs shortly thereafter when, in a scene mirroring the opening titles, the tape recorder is playing on top of a black desk with Cable's face reflected on its surface. Significantly, it is the same recording as the one played over the title credits, but with Cable's voice now heard responding to Bree's queries, and with Cable's motionless image contrasting with the speaking voice on the tape. At this point the detective story effectively becomes moot and the interplay between the voice and its power to control takes over as the central thrust of the narrative.

Two acoustic tropes are put in play by Pakula: the representational strategy of advancing the voice before revealing the speaker, and Cable's use of the tape recorder to capture and control the voices of his victims. At the heart of each of them is a basic function of creating a rift between the voice and the speaking body. According to Michel Chion, this is a significant operation because it replays the moment of the formation of the human subject and Jacques Lacan's theory of the access to language. Lacan notes that within this action there is an inherent violence and disruption in gaining access to the symbolic system of language because individuals must recognize that they are in fact the subjects of a larger symbolic order. Chion expands on this notion by advancing the idea that the slip between the speaking voice and the nonspeaking body relates to this moment before the formation of the individual and the accession to the symbolic realm of language. The act of reconstituting the voice and body, what

Chion calls "*mise-en-corps*," replays this moment of self-awareness, but it also leaves a remainder within the text: "But in truth, what we have here is an entirely *structural operation* (related to the structuring of the subject in language) of grafting the nonlocalized voice onto a particular body that is assigned symbolically to the voice as its source. This operation leaves a scar, and the talking film marks the place of that scar, since by presenting itself as a reconstituted totality, it places all the greater emphasis on the original noncoincidence."[23] This basic lie behind cinematic construction is precisely what rests at the heart of Pakula's film, and he uses this "scar" to represent the violence that is never manifested visually in the film. In its place, the violence is rendered on the sound track in the form of the recording, through which Cable repeats his acts of aggression by replaying it over and over again.

However, this technological representation of the control over the voice is contrasted by an entirely different form of control. Bree's character is seen throughout the film demonstrating her control over situations through the use of her voice, particularly as a means to solicit business and to comfort uneasy johns. Although the motivation behind this is complicated by her job as a prostitute, it is also manifested in her desire to be an actress, seen at several points during the film. Moreover, her move away from prostitution is emphasized in several key scenes where Bree discusses her life with a therapist (Vivian Nathan), and we see and hear the conversation from the therapist's perspective.[24] These scenes with the therapist are extremely important because they emphasize the power of talking to heal, and they also provide a strong counterpoint to the actions we see onscreen. In the sessions Bree demonstrates vulnerability and doubt in contrast to her headstrong attitude and control in the film. And as Robin Wood pointed out, the sessions take place in a narrative space somewhere outside the story space of the film. Even though all the sessions could be read as the same one—Bree's clothes never change during them—the development of her responses matches the development occurring in the central story.[25]

Pakula often chooses to use Bree's voice, reciting her fears and anxieties, to accompany images of her and Klute as their relationship develops. But, as Diane Giddis noted, a counterpoint between sound and image is often constructed when Bree's words belie her onscreen actions.[26] As the talks with her therapist continue throughout the film, Bree's voice eventually becomes entirely acousmatic when the brief introductory shots in the therapist's office are no longer used. In this way, the balance of power in the film is read through the access to the space of the acousmatic voice. Cable demonstrated his dominance by means of the recording apparatus in the beginning of the film, but as he starts to lose control, it is Bree's voice that accesses the privileged acoustic space. This is particularly intriguing, because unlike the unraveling of Cable's mental state as Klute grows closer to discovering the truth, Bree's voiceovers evaluate the onscreen events from some future perspective, commenting on the actions in the past tense. Not only does this place her beyond the telos of the narrative's trajectory, but it also gives her a position that transcends the binary relationship between Klute and Cable, one outside the narrative drive.

In this way, the entire narrative movement involved in *Klute* shifts Bree from a fixed position within the basic story mechanism to a place outside the narrative where she can assert her independence. Of course, this being a Hollywood thriller, there still needs to be a classical denouement to mark the closure of the investigation. This comes when Cable confronts Bree and forces her to listen to the tape of her friend Arlyn Page (Dorothy Tristan) being murdered. In one of the most remarkable scenes in the film, and an Oscar-winning performance by Fonda, Bree listens to the tape and breaks down emotionally in a single long take. The power of the scene is remarkable because, as Pakula recalled, "the idea of hearing a friend's murder seemed more terrifying in its sadism"[27] and it let the audience experience the terror of Arlyn's murder while watching Bree's reaction. It also mirrored the scenes of Bree at the therapist's office with the shots framed in exactly the same over-the-shoulder construction. The result is an effect of folding two narrative lines together, with one supplanting the other. If there is one fault that can be cited in relation to the development and growth of Bree's character in the film, it is the fact that at the moment when her agency is called for the most, Pakula chose to have Klute come to a last-minute rescue. After Cable tumbles out of the window to his death in slow motion, the scene dissolves to Bree's apartment where both she and Klute are packing up the contents, and Bree's voice is heard accompanying the shot in voiceover. This completes the shift within the story from the actantial dyad of Klute/Cable to Bree exclusively. Despite her physical presence in the diegesis, her voice has come to occupy the privileged acousmatic space, demonstrating both a control over her own life and that of the central speaking position in the film. The split between the image and the sound is complete as Bree states, "I'm going to miss him," at the same time as she and Klute are seen gathering up her bags to leave.

Oddly, this disjunction between sound and image allows Pakula to have his narrative cake and eat it too. Nearly every review cites the fact that Bree and Klute head off together at the end, presumably back to Pennsylvania and a "normal" life. After Bree's substantial character development throughout the film, this ending comes as a bit of a letdown, if not entirely studio-fabricated. However, because she is heard saying in voiceover to her therapist, "Maybe I'll come back—you'll probably see me next week," a final note of discord is introduced. Not only does it solidify a sense of independence and self-determination that had been built up throughout the film, but it also gives Bree a final speaking position, one located outside the framework of Klute's detection story. In this way the move of Bree's voice from being strictly joined to a speaking body to a completely acousmatic voice places her in a position of power in relation to the structural paradigm created by the film. Whereas Cable sought to control the women he terrorized through the apparatus of the tape recorder, Bree exerted her own control by using the power of her voice. Through the complex deployment of sound overlaps and advances, every main voice in the film but Bree's is granted the position of being acousmatic; but as the overlapping sounds diminish and she takes on the position of an acousmatic speaker, it marks a final return of power to Bree.

Pakula continued to explore the theme of surveillance in two other films in the 1970s: one reflecting the powerlessness of the protagonist in the face of totalizing conspiracy (*The Parallax View*) and the other the power of the individual to short-circuit the mechanics of the conspiracy (*All the President's Men*). Both films addressed a larger social schizophrenia that was introduced through the traumatic events of the 1960s. As pointed out by Stephen Paul Miller, "Surveillance functions as a means of reality-testing that assimilates the 'text' of the sixties into that of the seventies. Thus, in a sense, the mid-seventies closes the sixties."[28] In Pakula's paranoia trilogy, this collision between the public and private spheres is manifested through the configuration of the conspiracy as an agency unrenderable within the mise-en-scène. This is achieved through the optical strategies of long lens work, complex tracking shots to move from close to distant view, and abrupt shifts in tone through editing. However, neither of the later films displayed such a careful formal construction as the theme of emotional terrorism elaborated in *Klute* through sound techniques.

What followed in the wake of films like *M*A*S*H* and *Klute* was a series of films that explored the landscape of a post-political America through a chain of what Thomas Elsaesser identified as "the pathos of failure." Elsaesser, from the perspective of 1975, noted a dramatic shift in the status of narrative representation in American cinema. Unlike the traditional "antihero" configuration, which can be traced from films like *Little Caesar* (Mervyn LeRoy, 1931) and *Scarface* (Howard Hawks, 1932) through late-1960s variants *Bonnie and Clyde* (Arthur Penn, 1967) and *Easy Rider*, many of the films that appeared in the early 1970s bore the mark of the "unmotivated hero," existing in the narrative with the sole purpose of failure. As Elsaesser astutely states: "Clearly in a period of historical stasis, these movies reflect a significant ideological moment in American culture. One might call them films that dramatize *the end of history*, for what is a story, a motivated narrative (which such movies refuse to employ) other than an implicit recognition of the existence of history, at least in its formal dimension—of driving forces and determinants, of causes, conflicts, consequences and interaction?"[29] Elsaesser's assessment of the field of American cinema in the early 1970s reflects a trend whereby elements of political and social critique were often transposed through a shift away from the standard cause-and-effect chain of narrative. By breaking down the standard narrative conventions of character identification, the regular resolution of subplots, and the general teleological structure of classical Hollywood cinema, many of the films from the period sought to express meaning through means that were not contingent on the centrality of the narrative.

Two-Lane Blacktop (Monte Hellman, 1971)

The film that embodies the idea of the unmotivated hero and which uses sound techniques to convey the emotional states of its characters is Monte Hellman's *Two-Lane Blacktop*. The history behind *Two-Lane Blacktop* reveals it to be one of the most inimitable films produced by a major Hollywood studio. The film firmly resists both any linear sense of narrative cause and effect while also offering no conflict resolution or closure at its end. The film was shot nearly entirely in long take and deep focus, even in its multiple nighttime shots, and its minimal cutting strategy generally emphasizes a change of time or locale rather than building the typical patterns of shot/reverse shot within a sequence. With no major stars and very little dialogue, it is strange that such an atypical film ever got off the starting line. But the film was completed and immediately hailed as one of the most remarkable films of the year by numerous critics, even garnering the publication of its complete screenplay in the April 1971 issue of *Esquire* magazine. However, the film is more than just an oddity; its use of sound effects, speech, and music makes it a prime example of how sound is more than just an added-on effect and how it can be an equal partner with the images in conveying the story details.

It must be acknowledged that a film like *Two-Lane Blacktop* never could have been produced at a major studio, let alone as an independent production, if it were not for youth-oriented films such as *The Graduate* (Mike Nichols, 1967) or *Easy Rider*. The success of these earlier films was due in part to the recognition of an emergent youth audience interested in seeing their own concerns reflected in the story material, and the careful selection of popular recording artists on the sound tracks.[30] Both *The Graduate* and *Easy Rider* broke from the standard Hollywood model of using an instrumental score and a specially commissioned title song for crossover between the film studio and the recording industry. Instead, *The Graduate* used folk-rockers Simon and Garfunkel to provide lyrically based songs for the sound track with a few instrumental tracks. *Easy Rider*, on the other hand, used a number of previously recorded rock songs in place of a standard score. Originally the filmmakers commissioned Crosby,

Stills, and Nash to compose a sound track similar to that of Simon and Garfunkel's, but director Dennis Hopper chose to use the prerecorded music that was already in the temp-mix. The resulting sound track contained only one new performance, Roger McGuinn's "The Ballad of Easy Rider," a rewrite of Bob Dylan's "It's Alright Ma (I'm Only Bleeding)," which was used as the film's coda.[31] Part of the success of both films can be attributed to the cross-promotional marketing of the film music as a souvenir for spectators and as ancillary advertisements for the films themselves. This resulted in a number of rock sound tracks from *Zabriskie Point, The Strawberry Statement, Getting Straight*, and *R.P.M.*, each attempting to tap into the burgeoning youth market.[32]

The success of both *The Graduate* and *Easy Rider* led to a surge in the production of films targeting younger audiences, and many studios were willing to take gambles on new directors and new themes. After their breakthrough with *Easy Rider*, Bert Schneider, Bob Rafelson, and Steve Blauner's BBS Productions was offered a coproduction deal at Columbia with the single caveat that each of their films had to cost less than a million dollars.[33] In response to the BBS deal at Columbia, Ned Tanen, the newly appointed head of production at Universal, also secured a funding line for a series of inexpensive films made by a group of young directors. The first films produced by Tanen at Universal included *Diary of a Mad Housewife* (Frank Perry, 1970), *Taking Off* (Milos Forman, 1971), *The Last Movie* (Dennis Hopper, 1971), *Silent Running* (Douglas Trumbull, 1971), and *Two-Lane Blacktop*.[34] Though none of the films were box office successes, Tanen was eventually vindicated when his final pet project *American Graffiti* (George Lucas, 1973) returned a gross of over $55 million in 1973.[35] What the films did do is present a new model for small, personal projects made by fanatically devoted directors. In the case of Monte Hellman, he had originally been asked to direct *Two-Lane Blacktop* for Cinema Center, but after the film entered into pre-production, funding was withdrawn and Hellman had to shop the project around Hollywood.[36] After rejections from nearly every major studio, Tanen and Universal greenlighted the film, with Hellman taking a percentage in lieu of his normal director's salary.

Right from the start it was clear that *Two-Lane Blacktop* was not going to be the same as the flurry of other youth-oriented films that had flooded the market in the years since *The Graduate* and *Easy Rider*. One hint of this was the unusual casting choices made by Hellman and his team. To rewrite the script, Hellman brought in counter-cultural novelist Rudy Wurlitzer, whose 1969 novel *Nog* had greatly influenced the director. In the part of the two male protagonists Hellman cast singer-songwriter James Taylor and Dennis Wilson, the vocalist-drummer for The Beach Boys. The esoteric casting choices were not made because of the commercial viability of the musicians, but because they fit their character types perfectly. Taylor's wide-eyed stare concealed his recent heroin use, making him ideal for the troubled figure of The Driver, while Wilson spent his youth rebuilding cars for drag racing, making him an obvious choice for The Mechanic. But Hellman's choice to cast two major rock stars in the central roles and his decision not to have them sing anything in the film certainly must have struck the studio as unusual.[37] According to casting director Fred Roos, Hellman did not want either actor to perform his music in the movie because he wanted to avoid any "detrimental empathy" toward the characters.[38] The two other principals were cast shortly thereafter with nonprofessional actress/model Laurie Bird as The Girl and Warren Oates, a long-time veteran of Hellman's films, as GTO. This combination of predominantly nonprofessional actors taking on the roles of their generically named characters is just one indication of how the film was vastly different from most Hollywood productions of the time.[39]

Significantly, the film did not have a score and, like several of its predecessors, it featured a number of prerecorded pop songs. However, there is a major difference between the use of prerecorded songs in either *The Graduate* or *Easy Rider* and in *Two-Lane Blacktop*. The former films foregrounded music to the point where it would overtake all the diegetic sounds on the sound track, operating the same way as nondiegetic score music; yet the latter always positioned the music in relation to its diegetic source, paying strict attention to the sound's spatial signature. Despite the presence of a number of well-known pop songs (The Doors' "Moonlight Drive," Ray Charles's "Hit the Road Jack," and Kris Kristofferson's "Me and Bobby McGee"), no song is ever heard in its entirety, nor is it privileged in relation to the scene where it appears or its narrative content. For example, while it may have been possible to draw parallels between a song's lyrics and certain narrative content, Hellman refused such simple empathetic approaches and chose to use the music strictly as realistic background for the locations. Therefore, when GTO plays a prerecorded cassette of "Me and Bobby McGee" on his car stereo, the effect is simply that he was making a choice of music in an attempt to persuade The Girl to ride in his car with him. The content of the music is less important than a number of other cultural factors, namely the fact that his car is one of the first commercial vehicles to be equipped with a mobile cassette deck.[40] Similarly, there is no resonance between the music and the locations in the film other than the songs being germane to the milieu in which they are heard playing.

According to scholar Adam Webb, this was one of the choices that Hellman made to undermine audience expectations by using "the music as a disposable soundtrack: a distraction to which most of the characters are ambivalent."[41] Throughout *Two-Lane Blacktop* the music is meant to stay strictly in the background. Never heard nondiegetically, the music always carries the spatial signature of the location in which it is being played: car tape deck, AM transistor radio, bar PA, and so on. Despite the use of several well-known rock and pop songs, Hellman also used a large number of relatively obscure songs that can barely be heard above the din of the racecars and road noises. His musical choices were not designed to sell a sound track as an ancillary product; in fact, the licensing agreement from the Universal Studios' music marketing department did not even include a clause for a sound track release.[42] Instead, the music functions as just another element within an impressively polyphonic sound track.

This highlights another major choice by Hellman in the making of the film: a decision to have as little dialogue as possible and to let other types of sound tell most of the story. When pressed about the non-articulation of the "love story" in the film and his laconic central characters, Hellman discussed the absence of narratively motivated dialogue in an interview published in *Take One* magazine: "I always saw the film as taking place outside the dialogue. The dialogue has nothing to do with the film. The story of the film is entirely visual, and the dialogue is merely functional in terms of their everyday activities, and it's probably unique in that respect, because I can't remember ever seeing a film that didn't deal dialogue-wise with the subject of the film."[43] This was a conscious decision to minimize the amount of dialogue in the film in order to let the audience experience the effect of time on the road directly without interference or mediation. According to Aljean Harmetz, Hellman "was 'shocked' and 'amazed' to discover how wordless the film is. In some 10-minute reels, there are only two minutes of dialogue."[44]

To construct realism on the road, several versions of the two cars were used, each outfitted in different permutations to allow shooting both inside and outside of the cars. The entire film was shot on location and almost entirely in sequence when the cast and crew traveled from Los Angeles to Tennessee in the fall of 1970. To match the verisimilitude of the visuals, all the sounds in the film were recorded on location and no dialogue was looped in post-production. Not only did this mean a relatively complicated job of ensuring that the dialogue tracks were intelligible, but it also meant that a number of acting strategies had to be adjusted to match the aesthetic. Often the actors needed to shout their lines over the noise of the engine and the road in order to be heard on the recording as well as by the other actors. Efforts could have been made to soundproof the car or to shoot key dialogue scenes in the studio, but the sense of urgency and realism that is delivered in the line readings says more about the characters than their words do. The combination of dialogue and car sounds is essential to displaying how the central characters are comfortable only when they are on the road and foregrounds the existential difficulties of communication.

To achieve this balance between automotive sounds and dialogue, the sound mixer Charles Knight worked with his boom operator to equip the cars with a system of lavalier and ambient microphones. This provided a recording

FIGURE 8. The Driver (James Taylor) and The Mechanic (Dennis Wilson) listen attentively to the engine of the '55 Chevy after the first drag race in *Two-Lane Blacktop* (Monte Hellman, 1971).

that kept the dialogue and ambient sounds mixed together on the same track. According to the film's associate producer Gary Kurtz, the technique was done "to get the real feel of the shifting gears and engine sounds inside the car."[45] This was especially important because the combination of voice and automotive sounds gave a sense of veracity to key scenes throughout the film (see figure 8). In particular, the physical jarring of the shifting gears and changing speeds could be heard in the dialogue recordings. During one scene as the Chevy pulls into Tucumcari, New Mexico, on a rough road surface, the vibrations of the car are heard in the quaver of the actors' voices. In another scene, the sound of the wind becomes so loud that The Mechanic has to close his window to hear what The Girl is saying in the backseat. Any normal Hollywood production would have scrapped the shot and substituted another take with the window shut and soundproofed. However, by combining the dialogue recordings with ambient sound effects, Hellman was able to directly relate the visceral experience of the cross-country drive to the spectators.

One of the most distinctive aspects about *Two-Lane Blacktop* is revealed during the final credit sequence when both the 1955 Chevy and 1970 GTO receive featured billing alongside the actors. Hellman said in interviews that he wanted the cars to be characters just as much as the four principals. This was accomplished by the close attention given to both the visual presentation of the vehicles and the astute care used in creating their sound effects. As previously mentioned, the cars were equipped with external rigging that allowed for cameras to be placed at a variety of vantage points around and within the vehicles. Also, the use of the Techniscope process, a procedure that emulated CinemaScope by using half the normal space on the negative (two sprockets) per image, allowed the on-board cameras to run for twice the normal length of time per reel. Therefore, Hellman could "direct" scenes that played out with only the actors in the car and the sound recordist huddled on the floor in the space beneath the back seat. Many of the scenes were contingent on the audience feeling both the speed and the duration of the characters' travel without any dialogue, which was achieved through the careful recording and editing of the automotive sound effects. While a number of the car sounds were wedded to the dialogue in the production recordings, there was still a need for many specialized sounds that absolutely could not have been drawn from a stock sound library. To acquire these sounds, the task fell to Gary Kurtz, who worked directly with the sound team to assemble a compendium of specialized effects.

Kurtz received his cinematic training at UCLA and worked as a sound recordist, editor, and cinematographer for Roger Corman in the early 1960s. While working for Corman, Kurtz met Hellman and served as his associate producer on both *The Shooting* (1967) and *Ride in the Whirlwind* (1967). Kurtz was particularly attuned to Hellman's esoteric style, and his background as a sound recordist made him the perfect individual to supervise the collection of specialized automobile sounds for *Two-Lane Blacktop*. Because the cars were as important to the film as the characters, the sound effects had to elicit a sense of verisimilitude while engaging the perception of the spectator the same way that they controlled the attention of The Driver and The Mechanic. Although this was not a film made for "gearheads," it needed to achieve a certain stability in regard to the sound of the cars and the road. After the end of principal shooting and during the crew's drive back to Hollywood from the East Coast, Kurtz and the sound team made extensive recordings of the vehicles to provide a complete range of sound effects. To re-create the sound of the drag races, Kurtz and his crew took a number of cars down to an abandoned airfield in the Mojave Desert where they recorded hundreds of "run-bys" from both outside and inside the cars.[46] This gave Hellman and the sound editorial team at Edit-Rite, Inc. a wide assortment of effects choices that perfectly matched the cars, but also served to evoke the attentiveness of The Driver and The Mechanic to every aspect of the Chevy's performance.

This theme is expressed from the very beginning of the film as the revving of engines is heard over the Universal logo in place of the traditional studio fanfare. The deafening rumble of the cars raised the ire of Universal executives, who were afraid that projectionists were going to mistake the sound of the engines for a defect in the print.[47] In many ways this lack of the studio's authorial "voice" emphasized a shift in the control of the film—away from the studio and toward Hellman's personal vision. The sound advance of revving motors established *Two-Lane Blacktop* as unlike any previous films from the studio and emphasized that the use of sound effects was going to play a large part in the structure of the film. In fact, the entire pre-credit sequence was filmed without any discernible dialogue to guide the audience or introduce the characters. After the drag race is broken up by police, the title credits scroll by against a top-shot of speeding blacktop and double-yellow divider, with the sounds of the Chevy's engine, tires, and AM radio being tuned from station to station. This highly evocative sequence emulates the general effect of the film and cues the audience to concentrate on the diegetic sounds instead of relying on just dialogue or images. This is a point doubly emphasized when the first lines spoken by The Driver and The Mechanic are about the sounds of the engine and how they actively listen to ensure that the car is working at its top performance level. Because little or no backstory is relayed in the dialogue, the audience is forced to pay close attention to the sound track to discern the changing relationships between the characters.

A number of pivotal moments in relation to character growth are expressed directly through sound without any dialogue assistance. The first liaison between The Mechanic and The Girl is shown in a touching scene in Santa Fe where they silently change the sheets on the hotel bed with a tacit understanding of each other's intentions. When The Driver returns from a local bar, he hears the other two through the door of the hotel room and chooses to remain outside instead of interrupting their tryst. By point of comparison, the character of GTO is heard talking continuously throughout the film; yet every time he mentions details of his past, he tells a different story, casting into doubt whether anything he says can be taken as true. More than just a lack of dialogue, several of the scenes rely on a familiarity with the sound of the cars and their relationship to the characters as part of the narrative. In a scene where GTO has left with The Girl, The Driver betrays his emotional frustration by grinding the gears and accelerating to dangerous speeds. From the images alone it is very difficult to discern the speed of the vehicle because the scene is shot as a long take through the rear window of the Chevy. But the automotive sounds on the sound track emphasize the emotional panic and the resultant speed that propels the character to find The Girl. A scene that follows emphasizes this intrinsic connection between all the characters when GTO reacts disparagingly after hearing, but not yet seeing, the Chevy pull up at a roadside diner.

The careful deployment of sound effects in the mix allows the audience to perceive the journey the same way the characters experience it. Not only is the near ubiquitous presence of the automotive sounds significant, but it also opens up the potential for the suspension of sounds in the final drag-racing scene. Emphasizing the open-ended nature of the story and the aimlessness of the characters, Hellman refused simple closure by literally letting the film grind to a halt. To augment this effect, when The Driver prepares for the final drag race the individual sounds on the sound track are removed as the film shifts into slow motion. Finally, only the sound of the window closing is heard and the race begins in absolute silence. In a nice reversal of expectations, the visuals continue to slow to an eventual freeze-frame as the ground noise of the projector increases in volume until the audience hears just a rush of white noise. At the end of the crescendo, the frozen image burns up in the projector gate as the sound abates and the silent end credits appear. This utter abnegation of the cinematic convention of resolving the story with "The End" or a musical postlude was Hellman's method for returning audiences from their intense immersion in the diegetic world: "I wanted to bring the audience outside the film and back into the theatre. The film deals with time and speed and I wanted to add another dimension of time which is the time it takes for a film to run through a projector, and as the last image appears on the screen, the projector appears to slow down and stop, so suddenly we are dealing with time in another way. . . . It's a way of finally forcing the audience to come back to themselves and leave the film."[48] During this final scene of extreme character identification, where Hellman presents the drag race from The Driver's visual point of view for the first time, we are forced to sever that identification when the film refuses to continue. Although this may leave some viewers with a lack of resolution, the *in medias res* ending at a drag race mirrors the drag race heard over the opening studio credits. Hellman's decision to end the film with just the image lets it fold back onto the sounds from the beginning, dovetailing the film and creating an endless loop of wandering characters and unmotivated heroes.

Two-Lane Blacktop stands out against the other films released in 1971 as one of the three most original films to be made within the studio system that year. As with its counterparts *Klute* and *M*A*S*H*, the interplay between sound and image is central to the construction of its story and its narrative technique. In many ways 1971 is a high-water mark in the history of creative use of film sound in Hollywood. In part this is due to the unprecedented number of young filmmakers working in the commercial film industry, but it also has to do with the shifting roles within the production and post-production practices. Virtually no other producer would have devoted so much time and care to the gathering and use of sound effects as Gary Kurtz, and not until the late 1960s could a director move away from the studios to shoot exclusively on location. The collaboration of Hellman and Kurtz proved to be a valuable asset in the realization of the film's overall sound, and it was a lesson that Kurtz would take with him in producing two of the decade's other significant film sound achievements: *American Graffiti* and *Star Wars* (George Lucas, 1977).

THX 1138 (George Lucas, 1971)

The history of film sound evolution in the 1970s can be neatly traced in opposite directions by following the respective careers of George Lucas and Walter Murch.

In Lucas's case, the use of Dolby Stereo on *Star Wars* proved that technology, like action figures and toys, could be sold to audiences as an added feature of the onscreen spectacle. This, of course, led to the wide acceptance of Dolby Stereo as an exhibition format without a real trial period to demonstrate the technology's aesthetic potentials and shortcomings.[49] Yet in Walter Murch's case, sound was regularly marshaled to support the central story through whatever means were appropriate. For Murch, the skillful deployment of sound in a motion picture did not come by foregrounding the sound work; rather, "The challenge seemed to be to somehow find a balance point where there were enough interesting sounds to add meaning and help tell the story, but not so many that they overwhelmed each other."[50] In this way Murch was able to forge a creative alliance between the director and the post-production sound team to provide a link between designing the aesthetics of a film's sound and the conceptual goals of the director. This led to a number of powerful and evocative films that fully met the potential of creating cinema as an audiovisual art form.

The curious fact about Lucas's and Murch's divergent careers is that they both started by working together on two of the most intriguing films of the early 1970s: *THX 1138* and *American Graffiti*. Even though they both came from the film program at USC, Lucas and Murch met when the former was making a documentary about the filming of Coppola's *The Rain People* (1969) while the latter was mixing the film. Because Lucas and Murch felt a close alliance with the director, they both moved their families to the San Francisco area and helped to start American Zoetrope following Coppola's suggestion in early 1969 that a group of filmmakers relocate there to form a new cooperative studio. Not only did this free the filmmakers from the watchful eye of the major Hollywood studios, but it also allowed a measure of freedom from the draconian labor regulations controlling film production in Los Angeles.[51] When American Zoetrope embarked on its first production in late 1969, a feature-length remake of Lucas's award-winning student film *Electronic Labyrinth: THX 1138 4EB*, they found that a small, flexible crew allowed them to make the film more efficiently and quickly. Because the film community in San Francisco previously had been limited to commercial and industrial films, the local branch of IATSE was extremely cooperative when American Zoetrope drew its staff from the pool of untested film technicians rather than importing crews from Los Angeles.[52] To accommodate the needs of the filmmakers, production and post-production personnel were able to perform any job within their area instead of being restricted to a single job description. This opened up the ability for individuals to perform several duties according to the exigencies of the production. In the case of Walter Murch, it freed him to concentrate on constructing film sound during the entire production process, not just in the post-production phase.

Although the plot of *THX 1138* was rather derivative, borrowing heavily from Aldous Huxley and George Orwell in its vision of a futuristic dystopia, Murch's approach to sound allowed for several thematic issues to come into focus. Specifically, he was interested in utilizing the film's sound track to "carry as much of the film as the story, or the visuals."[53] Murch mentioned that the film was to present a world that was uncannily familiar in its use of present technologies. Instead of using synthesized sounds, Murch's job was to create a set of sounds from scratch that could evoke this "used future."[54] One method was by distorting regular sounds and voices to the point of estrangement, retaining the texture of the human voice while adding an unfamiliar electronic halo to the words. To achieve this, Murch created sideband distortion on certain dialogue recordings by broadcasting the production tracks by ham radio and re-recording them with the interference of the tuning.[55] Although the sounds were recognizable as voices, the technique placed the emphasis on the transmissive medium more than on the clarity of the dialogue, making the sound similar to the distorted voices of astronauts being broadcast back to Earth. But more than just creating a cacophony of voices and distortion, Murch was careful to create an acoustic space where the lines of undistorted dialogue could cut through the background sounds and distorted voices. This meant that the audience could be asked to strain to hear what the distorted voices were saying, but as soon as an undistorted voice was heard they would immediately latch onto that as the central carrier of narrative information. By creating this shifting figure-ground situation, Murch was experimenting with one of the fundamental aspects of film sound usage: the centrality of the speaking voice in narrative cinema. Although he does not directly articulate his strategies in regard to the manipulation of the voice, *THX 1138* works as an interesting experiment in how audiences have become conditioned to expect a particular intelligibility of the speaking voice in cinema. What Murch was doing was deferring those expectations to force the audience into a different listening relationship to the central story, a relationship that demanded more attention from the audience to make sense of both the linguistic significance of the voice and the hierarchies of speaking voices in the film.

This experimentation also extended to the level of sound effect manipulation. It was on *THX 1138* that Walter Murch started to develop a number of the theories regarding sound that would inflect his work on most of his future films. In particular, his work on the creation of metallic footsteps for the robotic police officers led to an epiphany regarding his "Law of Two-and-a-Half." The robots, played by actors in leather police uniforms and silver masks, were supposed to weigh hundreds of pounds and Murch created the sounds to match the intended concept. "[In] the film we wanted them to sound massive, so I built some special metal shoes, fitted with springs and iron plates, and went to the

Museum of Natural History in San Francisco at 2 A.M., put them on and recorded lots of separate 'walk-bys' in different sonic environments, stalking around like some kind of Frankenstein's monster."[56] But a difficulty arose when trying to edit in the footsteps to match the walking robots; the complexities of synchronizing the footsteps made it nearly impossible to make the scene sound realistic. According to Murch, "It was kind of daunting because you have to sit there cutting every footstep in, but to my relief I found there's a certain threshold where the mind simply rolls over and allows anything to happen, and that point seems to occur between the numbers two and three."[57] The psychoacoustic effect of the "Law of Two-and-a-Half" is that when sounds are heard in an aggregate lump the mind is able to discern synch for two similar sounds in a group; when that threshold is passed, there is a perceptual connection made regardless of exact synchronization. This was the first of Murch's many discoveries regarding the nature of film sound while working on Lucas's films, and one that would serve him well in future films, particularly in the massive sound effect work involved in *Apocalypse Now*.

One reason why Murch was able to experiment with the use of sound in Lucas's films was the flexibility of the work situation in regard to recording and mixing sounds. Unlike the strict hierarchical divisions of labor enforced in Hollywood, the use of San Francisco as a production site gave Murch the freedom to operate in any number of post-production roles without fear of union backlash. Another distinct advantage was the equipment available to him at American Zoetrope. Coppola was one of the first directors to embrace the use of new technologies, and in the early 1970s the gap between professional and consumer audio equipment was fading rapidly. One advantage that Coppola discovered while making *The Rain People* was that entire films could be edited and mixed efficiently and quickly using KEM flatbed editing tables. According to Murch, "In fact, [the gap] faded to the extent that it now became technically possible for one person to achieve what several had before, and the other frontier between sound-effects creation and mixing also began to disappear."[58] Of course, for this to develop there needed to be an understanding between the director and his sound team that the film's sound was going to play an integral part in the story development. To achieve this and to maintain a system of "internal logic" to the sound effects, Lucas had Murch cut and edit the sound track simultaneously with the creation of the visual rough cut. According to Murch: "The work on the sound effects and music was begun during shooting and carried out during the editing, so that the sound and the picture grew together and influenced each other. By the time there was a rough cut, we had also mixed a 'rough cut' sound track to go with it."[59] The actual simultaneous construction of a sound track along with the picture edit was something that was unheard of in standard Hollywood practices, but it became one of the major advances enacted in the collaborative filmmaking approach of both George Lucas and Francis Ford Coppola.

Because Murch was free to perform a "one-man mix," a process that was simultaneously developing on the East Coast with sound mixers Dick Vorisek and Richard Portman, it meant that he could control not only what sounds were utilized in the film, but also how they were presented in the final mix. Murch chose to name this procedure "sound montage," partially to reflect the way that it reflected the process of film editing and montage, but also in part as a way to deflect any union criticism of him working as both a sound editor and a re-recording mixer. According to authors Lynda Myles and Michael Pye, the relatively unorthodox method of recording and mixing sounds in conjunction with the cutting of the film "allowed the intricate sound track to grow along with the images that Lucas himself was photographing, directing, and editing. The sound montage was an organic part of the film, not a decoration imposed afterward."[60] This organic unity between the sound track and the images is the trademark of all of Murch's film work and one that distinguishes his films from those of others because of the conceptual link between Murch's sound theory and his correspondent sound practices.

Although both Murch and Lucas had made great strides in the advancement of both the rethinking of sound labor during the production process and the integration of sound techniques with the needs of the story, most audiences were shocked by the film's radical audiovisual style upon its initial release in March 1971. Warner Bros., the studio that funded Coppola's American Zoetrope venture, was offended by the content and representational form of *THX 1138* and reedited the film for its theatrical release. Earning only $945,000 against its production cost of $777,000, the film not only was a flop for the studio but also prompted them to rescind their funding of both American Zoetrope and any scheduled Coppola productions.[61] Although this was a major setback for both Lucas and Coppola, the latter already in the pre-production phase of his magnum opus *Apocalypse Now* (1979), it did not deter the filmmakers from collaborating or from following their own idiosyncratic visions. Two years later, Lucas had secured funding through Ned Tanen at Universal to make *American Graffiti* with Coppola signed on as executive producer.

The development of new sound aesthetics at the start of the 1970s matched the visions of a generation of American filmmakers who would move from the independent margins to the center of mainstream commercial cinema. Even though Altman, Pakula, Coppola, and Lucas all benefitted from increased budgets and larger audiences in the 1970s, they each continued to explore and refine sound techniques in the aid of storytelling. In particular, Altman expanded his use of offscreen sound as commentary and combined it with a collaborative sound method borrowed from Arthur Penn's work in the late

1960s. The resulting "democratic sound" found its apotheosis with the multi-track dialogue recordings on *California Split* and *Nashville* that let Altman "un-mix" the dialogue in post-production and redeploy the voices across the soundscape. Coppola continued the experiments in subjective sound started by the British New Wave directors in films like *The Rain People*, refined his techniques in the *Godfather* films (1972, 1974), and made them the heart of the story in *The Conversation* (1974). And borrowing the use of source music from *Two-Lane Blacktop*, Martin Scorsese turned the use of popular music as a compilation score into a two-tiered system of signification in *Mean Streets* (1973), and continued to refine the use of both score and source music in his films to develop a sense of subjectivity through music in his narratives.

PART TWO

Director Case Studies (1968–1976)

5

Francis Ford Coppola's American Zoetrope and Collective Filmmaking

Technology, this incredible stuff we're dealing with, requires balanced human beings to embrace it and function with it in a positive way.

–Francis Ford Coppola

In an effort to distance himself from the vicissitudes of filmmaking in Hollywood, Francis Ford Coppola set up his own studio, American Zoetrope, in San Francisco and launched a filmmaking collective in the process. With collaboration as its keynote and with the relaxed union regulations of the Bay Area, this led to a new method of filmmaking and dynamic experiments in sound techniques. Coppola embarked on his first independent production when he resurrected an old script treatment from his days at UCLA for *The Rain People* (1969). Frustrated with his experiences making the overly ambitious *You're a Big Boy Now* (1966) and the restrictions of filming *Finian's Rainbow* (1968) at the Warner Bros. Burbank Studio, he decided to embark on a cross-country trip to shoot *The Rain People* on location and in sequence. Inspired by the success of smaller-budget European films like Claude Lelouch's *A Man and a Woman* (1966), Coppola set out with a small crew, three equipment vans, and two station wagons to make his own personal film about restless housewife Natalie Ravenna and her flight from middle-class normal. Warner Bros. producer Kenny Hymans, pleased with Coppola's work on *Finian's Rainbow*, gave the director a shoestring budget of $800,000 and carte blanche to shoot the film on location.[1] An intimate character piece, the film rests on restrained performances by Shirley Knight as Natalie and James Caan and Robert Duvall as her would-be paramours. In keeping with the intimacy of the story, the film itself is surprisingly delicate in its sound work and resists many of the clichés of melodrama, most notably in its use of music.

The Rain People (1969)

The overall sound of the film is the result of contributions from four individuals: sound mixer Nat Boxer, boom operator James Sabat, sound editor Walter Murch, and composer Ronald Stein. Boxer and Sabat were relative neophytes working as freelancers out of New York and Murch was a recent graduate of the USC film program. Stein, on the other hand, was a veteran composer, having worked on over fifty low-budget films, many of them for American International Pictures including Coppola's directorial debut, *Dementia 13* (1963). The sound team ultimately crafted an intimate portrait of Natalie's complex and contradictory psychological state. Michael Pye and Lynda Myles noted: "Coppola had been able to pick his associates, and on *Rain People* he learned to trust them, to discuss things with them. On location he made take after take of single scenes, waiting to see what would emerge from the final shaping on the editing table; and the sound, although less central to the film's concerns than the track of *The Conversation*, was almost as complex and sophisticated. Coppola learned to put himself more at risk, to allow a team to develop his ideas."[2] This collective approach to film sound was the cornerstone of his working method, and the tension among competing sound practices in the film would eventually coalesce into Coppola's personal sound aesthetic.

The tension can be sensed right from the beginning of the film when the stinger chord that introduces the first image is replaced with the realistic sound of garbage men plying their trade along a rain-dappled suburban street. The sounds of the truck, the garbage men rattling cans, and crows cawing in the distance belie the fact that nearly all sync points are either hidden or offscreen. The truth is that Walter Murch recorded and edited all the sound effects in the film other than the dialogue, location sound effects, and Stein's score music. According to Murch, "There I was, in a cabin up Benedict Canyon, all alone with the film, a Nagra tape recorder, a Moviola, and a transfer machine, recording and adding the sound effects. I became a little paranoid—I was nonunion and an ex-film student working on a studio feature film—and I felt I couldn't go to any of the film libraries and get prerecorded sound."[3] As a result of Murch recording all the effects himself, there is a stylistic consistency to the range of sounds and their deployment in the picture to support the realism of the story. Diane Jacobs noted that "many critics have attacked *Rain People*'s style, its craftful dodging from realism (the garbage trucks sounding like garbage trucks, the *cinema verite* highway noise superimposed over Natalie's phone conversations) to glossy cinematography (close-ups of windshields, pannings around lakes and landscapes),"[4] but what she didn't mention is that it is precisely this tension that Coppola sought in the film. Whereas the images tend toward the artistically expressive, the sound stays anchored in a form of "audio vérité."

It is important to distinguish between "realism" and codes of realism in relation to sound use in *The Rain People*. The effect of audio vérité is to create a sense of realism between the sounds and images that evokes a documentary sensibility. While this realism is evident throughout the film, it is fabricated from disparate sonic elements. In the process Murch was relying on the codes of realism drawn from documentary filmmaking—simultaneity of image and effects, ever-present ambience, and a vast extension of sounds outside the frame[5]—for audiences to develop a direct attachment to the characters in the story. Although all three elements are also used in feature films to create a realistic diegesis, often ambience and external sound effects would be reduced or even eliminated to foreground the dialogue and music. In the case of *The Rain People*, Stein's music is used sparingly, almost exclusively for transitions and travel montages, and the dialogue is recorded in a way to preserve both the space in which the characters are speaking as well as the way the characters hear each other. But it is through sound effects and their metaphorical application that Murch was able to control much of the tenor and emotion in the film.

Returning to its beginning, the sounds of the garbage trucks carry over several focal shifts and cuts that lead the camera to a suburban house and the darkened bedroom of Natalie and Vinny Ravenna. The sound and levels of the garbage trucks do not change over the cuts to match the perceived move away from their source, and we hear their sounds as Natalie begins to stir and extricate herself from under Vinny's arm. Along with the Foley sounds of Natalie's sheets rustling are low-level inchoate voices, cheering and presumably saluting the guests of honor at a party. At this point it is unknown whether these sounds are offscreen, coming from an unseen radio or television, or internal, emanating from Natalie's mind. The camera cuts to Natalie in the shower and the white noise of the water creates a sonic shock cut that corresponds with the bright white lighting of the bathroom. Yet behind the shower sounds are still heard distant voices that hint that we are meant to interpret this verbal clamor as internal to Natalie. The typical convention would be to raise the levels of the voices and let them introduce a visualized flashback that illustrates what Natalie has been thinking. Instead, Coppola and Murch leave the voices at a subliminal level to entice the audience to develop their own understanding of the sequence.

The film's opening reflects Coppola's cultivation of a metaphorical approach to sound in the film, one that he develops more fully in his films from the 1970s. Even though there is no discernible dialogue, critics like Peter Cowie extrapolate that Natalie "stirs awake in a dark dawn after what sounds to have been a quarrel with her husband."[6] This is because Murch's audio vérité approach deployed sounds germane to the environment in a way that augmented and enhanced an understanding of the narrative. The visualization of the garbage trucks justifies the presence of their sound at the Ravennas', yet the sounds are just a little too present to match their distance from Natalie's bedroom. At the same time, the low-level voices could be from a diegetic source, but it is only several minutes later that they are revealed to be a memory/flashback of Natalie and Vinny's wedding. In the interim the montage of the voices and the truck sounds work to create the aural illusion of an argument that grounds Cowie's interpretation of the scene.

Murch was able to construct several moments of metaphorical sound use because Coppola left the story deliberately ambiguous and did not provide a backstory to explain why Natalie would leave her husband so abruptly. Correspondingly, the audience has to search for other clues to explain her motivations and mindset. When she finally stops running away and checks into a hotel in the middle of Pennsylvania, Coppola lets the scene play out wordlessly as Natalie makes a cup of coffee and then settles into bed. For a full forty seconds Natalie is seen sitting on the edge of the bed waiting for the water to boil, accompanied only by the rumbling traffic on the nearby highway and sounds bleeding through from other rooms. Toilet flushes, showers, footsteps, muffled conversations, and whistling are all heard through the walls with none of the sources being visualized. At the same time that it enhances the diegetic realism, her inability to block out the sounds hints at her internal mental turmoil. Only after she decides to try to sleep does Coppola choose to visualize her thoughts by showing flashes of an earlier hotel tryst with Vinny, yet we hear none of the sounds from the flashback, just the monotone drone of the highway.

Key to the use of sound in *The Rain People* is how the film shifts subtly from objective realism to Natalie's subjectivity through sound. Another way this occurs is through the choices made by location sound recordists Nat Boxer and James Sabat. Boxer and Sabat became purveyors of what was known as the "East Coast Sound," pioneered by New York–based sound recordists and boom operators in the 1960s and 1970s. Even today there remains a sense of communication between the director, the production recordists, and the post-production staff in New York that is distinctly different from the insularity and compartmentalized workflow in Los Angeles. Boxer started in the film industry in the 1960s on films like Robert Wise's *West Side Story* (1961) and Sidney Lumet's *Fail Safe* (1964) before he met Coppola working on *You're a Big Boy Now*. With the exception of *The Godfather*, Boxer continued to work with Coppola on all his films through *Apocalypse Now* (1979), for which Boxer ran his sound team from the position of a boom operator rather than as a sound mixer. This approach came from Boxer's education at the New School for Social Research film program, where he recognized the importance of developing a boom microphone strategy that accommodates the presence of the set lighting.[7] When Coppola asked him to head the production sound team on *The Rain People*, Boxer decided that he would prefer to work as the boom operator, leaving the intricacies of mixing to his colleague James Sabat. Working closely with cinematographer Bill Butler, Boxer established a rapport that allowed him to integrate the microphone strategies into the film's complex lighting and cinematography.

In the strictly segmented workforce of Los Angeles, the idea of a lower-ranking boom operator telling a production mixer what to do would be unheard of. But in New York, not only was Boxer's contribution recognized, it was regularly sought out. Boxer was a proponent of the use of sound perspective in his recordings, which was controlled by his skillful operation of the boom. Boxer recognized that the shift of control from sound mixer to boom operator presented several possibilities:

> If the boomman is the head soundman, he has the power of making a decision. . . . When he says, "Can I see the frame line because I want to put the mike right outside?" he gets a response. So he is making a decision which is first-hand and direct. Now flying that microphone around and making it work so it doesn't disturb everybody is one of the boomman's functions. The other function is making the right choices: what microphone to use, where to place it, when to decide whether you are going to go underneath the shot and boom from below or boom from overhead, are you going to hide additional microphones, are you going to use a lavalier or a radio mic or both?[8]

This process of moment-to-moment decision making reflects a performative approach to sound recording that required the boom operator to function almost as an additional cast member, needing to be blocked during the rehearsals along with the actors. By this method, Boxer was able to get the best possible sound while adjusting his boom work to the changing demands of the production.

The freedom for Boxer to operate as the main sound person on set was only possible due to the relaxed system outside Los Angeles. The close bonds established with the directors and cinematographers were a direct byproduct of this relaxed production approach:

> On the East Coast the industry is set up more on a free-lance basis, so if a technician doesn't like the studio or the studio fires him, it's not the end of the world—there are still plenty of other producers to work with. On the West Coast there are really only the major studios, and if you alienate one you've eliminated just about 30% of the work area—or maybe more if they speak to someone else. Also, on the West Coast there are many more social divisions: the director is the General and you almost never speak to the director. . . . On the East Coast that is unheard of.[9]

This direct communication between Boxer and Coppola enabled him to create a number of carefully recorded production tracks that balance the demands of dialogue recording, cinematography and lighting, and the narrative demands of the story. Boxer also estimated that this resulted in less than ten percent of the lines on his film being looped. By blending a radio microphone feed with that of a simultaneous boom mike, Boxer regularly eliminated set noise while adding some of the boom feed to create a sense of spatial perspective.[10] Sabat mixed the channels live while Boxer controlled the reverberation by adjusting the location of the boom mike. In this way he was able to achieve three distinct goals: a clear recording of the dialogue in a noisy situation, the inclusion of sound perspective in the recording, and the wedding of the two elements on a single track, making it easier to edit in post-production.

At the center of *The Rain People* are three extended dialogues between Natalie and Vinny that occur over a pay phone. In each instance Boxer used a boom mike to record Natalie as she spoke with Vinny, who was recorded directly over the long-distance phone line. The traditional method for recording phone conversations was to use a dummy telephone and record one actor with a boom mike while the director or script supervisor read the lines for the second character. The process was then reversed, with a second actor in a different location reading his lines, and the two shots were edited together. But in *The Rain People* Coppola chose to keep the camera on Natalie's character and never cut away to reveal Vinny (see figure 9). This constructs a curious effect, where "his presence, though unseen, is crucial to the film, serving as a kind of conscience reacting against Natalie's moments of self-pity," observed Peter Cowie. "Coppola and his sound engineer, Walter Murch, treat this voice with care, reducing it to an incomprehensible mutter at some junctures and magnifying it at others so it outweighs Natalie's."[11] Cowie is correct in saying that Vinny's acousmatic voice functions as the metaphorical voice of reason, but his physical absence from the scenes, and from the film in general, places his character firmly in the realm of the past and outside the events experienced by Natalie. Moreover, even though Murch may have adjusted the sound mix during post-production, the effect of Vinny's telephonic voice was created by Boxer and Sabat, who used a live telephone line to connect actors Shirley Knight and Robert Modica and mixed their voices to one single track on location. The result is a more dynamic exchange where the characters' voices freely interact and overlap, and the dialogue sounds like actual telephone conversations rather than Hollywood studio reconstructions.

Despite Coppola's innovations in sound, vision, and technique, *The Rain People* is still a flawed film in many ways. By leaving Natalie's backstory obscured there was little chance for character development or change to occur, and Coppola's avoidance of a conventional Hollywood ending made the film

FIGURE 9. Natalie (Shirley Knight) calls her husband long distance to explain why she left him in *The Rain People* (Francis Ford Coppola, 1969).

decidedly downbeat for mainstream studio fare. In addition, there are several scenes where Coppola is clearly referencing his favorite directors and films, though often without any sort of allusive resonance that would enhance the story being told. The interpolated flashbacks borrow heavily from Lindsay Anderson's *This Sporting Life* (1963) and Richard Lester's *Petulia* (1968), yet unlike both examples Coppola reveals very little through the flashbacks that was not already signaled though dialogue. And when Robert Duvall's character Gordon, a motorcycle policeman, is first introduced, the scene is constructed to mimic a similar scene in *Psycho* (Alfred Hitchcock, 1960) while Ronald Stein's score apes Bernard Herrmann's music just to underline the reference. In fact because of the attentive sound work throughout the film, Stein's score tends to intrude whenever it is present. The first seventeen minutes of the film unfold without any score music, and when Stein's orchestrations arrive, coincident with the use of helicopter shots to create a travel montage, suddenly the film lurches from its intimate subjective realism to emulate the stylistic excess of Lelouch's *A Man and a Woman*.

The Godfather (1972)

Coppola's experience on *The Rain People* solidified his confidence in the collaborative nature of filmmaking, and he continued to work with Murch and Boxer throughout the 1970s. For Coppola, his next film posed an especially complicated problem. Despite grand expectations from Paramount Pictures, Coppola still sought to tell a small, personal story about the Corleone family in his adaptation of Mario Puzo's best-selling novel *The Godfather*. Because the studio dictated many of the personnel working on the film, Coppola was somewhat limited in establishing his own stamp on the film's overall look and story. One of the ways he asserted his control was over the sound of the film. Coppola brought Walter Murch into the project as "post-production consultant," which allowed him to work as a freelance sound editor and mixer outside the purview of the Los Angeles sound unions. Murch, along with veteran re-recording mixers Bud Grenzbach and Richard Portman, was responsible for shaping the sound aesthetics of what has become perhaps the most written about film in history. Surprisingly, despite volumes dedicated to its storytelling, acting, and production history, there has been very little critical analysis of the film's sound.

In keeping with a story world that few people knew about at the time of the film's release, *The Godfather* begins and ends in the cloistered study of Don Corleone. The much-quoted first line by Amerigo Bonasera, "I believe in America," is delivered in a visually and acoustically muted environment. It is not only Bonasera's monologue that sets the narrative tone for the film, but also the chiaroscuro lighting patterns and the lack of ambient sound. This absence of sound is made even more noticeable as the camera slowly zooms back to reveal more of the room and eventually the head and shoulder of Don Corleone, while occasional creaks of the leather furniture introduce some of the basic room sounds. The audience soon discovers that this is the Long Island home of the Corleone family where, at that moment, the wedding of his daughter Connie is under way. But from the beginning, the hushed ambience and emphasis on Bonasera's broken English affects the audience in several ways. First, it sets the tone for the story of Italians living in America and the struggles they face in the postwar 1940s. Second, it constructs Don Corleone as an attentive, contemplative individual whose simple words belie the depth of his intellect. And third, as in *The Rain People*, the absence of ambient sounds when they should be present hints at the subjective nature of sound in the film. At several times in *The Godfather*, sounds are not just objective correlatives for the sources we see but reflect the way in which they are heard or not heard by particular characters.

Language and speech are two acoustic elements that any audience member notices about *The Godfather*. The fact that a large proportion of the film is spoken in Italian and Sicilian dialect immediately foregrounds the significance of language, and there are several times when individual words or even entire passages of dialogue are not translated. For English-language audiences this forces viewers to try to understand the context of the words and to study the speech patterns instead of just their semantic meaning. Sound theorist Michel Chion identified three different modes of listening when audiences watch films. Generally an audience engages with a film either through causal listening, where they search for the source of a particular sound, or through semantic listening, where meaning is sought in the coded nature of language.[12] But if the code of language cannot be deciphered, such as the use of unfamiliar foreign terms or untranslated dialogue, the listener often enters a third level: reduced listening. Chion referred to reduced listening as "the listening mode that focuses on the traits of the sound itself, independent of its cause and of its meaning."[13] It is common practice in *The Godfather* for the audience to enter into a mode of reduced listening to consider the qualities of sounds themselves as bearing their own meaning. This is at the heart of Coppola's metaphorical sound use and he encourages the audience to listen to the sounds in the film for more than their literal meanings.

In the case of the voice, much screen time is spent making the audience listen to not only what the characters are saying but how they say it. Arguably the classic example of this is Marlon Brando's voice as Vito Corleone. Like his extensive makeup that enhanced the aged quality of his character, he repurposed his voice to develop a gruff susurrant quality befitting the Don's age and experience. In *The Godfather* we don't know whether this voice is the result of some past accident or trauma, like the conspicuous dent in his forehead, or if it is the mark of a wearied and downtrodden old man. But more than just affect, his voice is a cultivated construct that establishes and reinforces his power. His hushed vocalizations make his interlocutors, as well as the audience, listen closely to his words and their delivery. Chion commented that "Don Corleone's husky, intimate voice makes the listener conscious of its presence and its timbre, as well as its fabricated quality. Ostentatiously reorganizing space around itself, it is a voice that obliges one to listen and that one is conscious of listening to. It is a quintessentially cinematic voice, since it exists solely in vocal closeup."[14] Part of the irony in Chion's comment is that what he refers to as "vocal closeup" does not mean that Brando was recorded differently than the other characters or that his lines are more pronounced; rather it is the effect of the voice on the audience such that we tend to amplify its volume and meaning because of the attentive listening patterns the film demands.

The classic example of the 1970s cinematic voice, Brando's voice for Don Corleone is just one of many significant speech and vocal patterns within *The Godfather* and its sequel. Voice and accent serve as elements of typage to immediately orient the characters for the audience. The heavy Italian accents from Bonasera, Enzo the baker, and Luca Brasi quickly establish their characters and the milieu at the start of the film, but equally important are the lack of accents heard from the second generation of Italian Americans, particularly the Don's sons Sonny (James Caan), Fredo (John Cazale), Michael (Al Pacino), and their adopted "German-Irish" brother, Tom Hagen (Robert Duvall). More than the accented quality of character voices, the control of the voice and who gets to speak also provide valuable information about the characters and their relationships. Tom, the Don's *consigliere*, is the most measured in his use of words, keeping with his legal expertise and signaling his strong sense of control. On the opposite end, both Fredo and Sonny are prone to outbursts, taken as a sign of weakness in the former and an inability to control the latter's temper. When Sonny interjects during a business meeting with members of the rival Tataglia family, he is upbraided by his father who comments, "They talk when they should listen." Michael, who becomes the main character across the *Godfather* saga, starts somewhere in between his brothers. During his interactions with his girlfriend Kay (Diane Keaton) he is loquacious as he describes the backstory of the Corleones, adding, "That's my family, Kay. That's not me." Yet as the film progresses he becomes more reserved and his voice hardens.

Michael's voice, like his father's, carries a great deal of information about his character, his thoughts, and his narrative progression. After his father is attacked by the Tataglia family at the behest of the up-and-coming drug kingpin Solozzo (Al Lettieri), Michael's voice begins to change. At first, this is just a surface characteristic of his reticence to engage in the family "business," but there are subtle hints that the change lies much deeper. When Kay calls to check on him, Michael refuses to return her "I love you," deflecting the sentiment by saying "I can't talk." This linguistic platitude becomes literalized when the headstrong Captain McCluskey (Sterling Hayden) breaks Michael's jaw outside the hospital where Don Corleone recovers from gunshot wounds, and he spends much of the middle portion of the film with his jaw wired shut. More than a metaphor, Michael's broken jaw reflects a major internal change that is registered in his voice. Because speech is difficult and even painful, Michael learns to choose his words carefully. In the scene that follows, Sonny and Tom argue over the best strategy to deal with the Tataglias and Solozzo as the camera slowly tracks in on Michael, waiting for him to speak. Outlining his plan to assassinate both Solozzo and McCluskey, Michael's measured speech reflects a change in his personality, one that is confirmed when he completes the murders as proposed. The rest of the film finds Michael's voice hardening and becoming more hushed like that of his father as he assumes control of the family business.

Coppola showed a masterful sensitivity to the voice and its meaning throughout *The Godfather* and its sequel, and from a structural perspective the deployment of speech also bears significance. Like Sonny's hotheaded interjections, there is a hierarchy of who can speak and who can hear within the film. Vito's daughter, Connie (Talia Shire), chastises Sonny for bringing the details of the family business into the home: "Papa never talked business at the table and in front of the kids." Moreover, Sonny's verbal excess is contrasted by Luca Brasi's "quiet purposefulness."[15] Although there are several loud verbal skirmishes throughout the film, there are equally as many moments of intimate

FIGURE 10. The sound of finality after Michael (Al Pacino) becomes Don Corleone and a closing door excludes Kay (Diane Keaton) from his world in *The Godfather* (Francis Ford Coppola, 1972).

whispered conversations. When, at his father's funeral, Michael realizes that he has been betrayed by his loyal *caporegime* Tessio, he narrates the events that will unfold to Tom in a hushed exchange. Similarly, in *The Godfather Part II*, when Michael discovers that he was betrayed by his brother Fredo, his accusation, "I know it was you, Fredo. You broke my heart," is much more powerful when whispered directly into Fredo's ear. Yet, unlike his father, Michael's inability to control the modulation of his voice hints at their primary difference and Michael's growing hubris. Vito, after returning home from the hospital, queries Tom about Michael's whereabouts by summoning his energy to whisper the question. After discovering that Michael was sent off to hide in Sicily, the Don reacts with silent tears. When, at the end of the first film, Kay confronts Michael about whether he engineered the killing of Connie's abusive husband, Carlo, he erupts: "Don't ask me about my business, Kay!" Then, after she inquires again in a barely audible voice, Michael mollifies her with an understated "no." Michael's final lie is reinforced by Clemenza swearing his loyalty as the study door is closed, thereby separating Kay from Michael's business and underscoring his ascension as the new Don Corleone (see figure 10).

Perhaps the most stirring thing about a film about the Mafia and gangland activities is how rarely the film resorts to the clichés of the gangster genre. Different from the stereotypical mob bosses, the Corleones don't maintain their control simply through the brute force of a gun. Instead, their power is more insidious, and the sounds of violence in *The Godfather* are downplayed when expected according to genre conventions only to surprise the viewer later in their unexpected ferocity. Even small sound effects, like the clear trip of the door latch when Kay is shut out at the end of the film, carry a weight of significance. According to Walter Murch: "The door-close . . . has to be true to what we perceive objectively: the physicality of the door and the space around it. But it also has to be true to the metaphorical impact of that door-close, which is, 'I'm not going to talk about my business, Kay.' That *ka-lunk*, that articulated sound of solidity, has to express something of the finality of the decision."[16]

Many of the sounds in the film function on two levels; first, reinforcing the realism of the diegesis, and second, providing a metaphorical meaning that resonates with the film's larger themes. Film critic Judith Vogelsang identified several sounds tropes in the film that she adumbrated in her essay "Motifs of Image and Sound in *The Godfather*." Specifically, Vogelsang identified the repeated use of thunder, loud echoes, and screams that preface moments of violence. On the surface these effects can be linked to their stereotypical use in thrillers and horror films, yet Coppola mobilizes them for different purposes. It can be argued that the rumbles of thunder that are heard before the discovery of the horse's head in Jack Woltz's bed or the first appearance of McCluskey are acoustic clichés used as harbingers of danger. However, the use of echoes and screams in the film take on different levels of meaning.

Reverberation, the literal sound of a space, is one of the most powerful markers of realism available to filmmakers. The single omnidirectional microphone sound heard in the early films from Jean-Luc Godard provided a verisimilitude that matched the visualized spaces. Even though it could obscure the dialogue among the other sounds present on location, Godard preferred this method to reinforce the realism of the scenes. By contrast, most American films favored a technique of closely miking dialogue without regard to the distance of the speaking character from the camera. Very often this emphasis on dialogue intelligibility meant eliminating reverberation and sacrificing any sense of sound space.[17] This effect can be heard in the dialogue scenes between Tom Hagen and Woltz as they walk across the film producer's garden in long shot but are heard directly. Yet there are several other moments in the film where the sounds of spaces take on their own meaning.

The first such instance occurs when Luca Brasi (Lenny Montana) tries to infiltrate the Tataglia family and agrees to meet Bruno Tataglia at his club. The film follows Luca, strolling through the long hallway of a building, for a full forty seconds before he arrives at the bar. As he walks through the space, his footsteps reverberate along the tiled hallways and the sound of whistling is heard in the distance. Although this can be read as supporting the general realism of the period, where his creaking shoe leather is heard quite clearly, the overt presence of these sounds and the reverberation of the hallway take on a portentous quality. This effect occurs again when Michael goes to the hospital to visit his father only to find him unattended. As he moves down the hallway, a distant, repeating sound is heard reverberating. Only when he draws closer to the workers' lounge is it revealed to be a record player with the skipping needle repeating the word "tonight." Realizing that his father is unguarded, Michael and the night nurse move his bed to another room, taking care to make sure that others do not hear their sounds. Suddenly the sound of the front door banging reaches them, followed by slow echoing footsteps methodically moving closer. Michael listens, like us, for clues to discern the nature of the threat, only to discover that the interloper is Enzo the baker. As a result, each subsequent instance that heavy reverberation is heard in the film, earlier instances are summoned to mind and the audience is conditioned to expect the possibility of imminent danger.

Vogelsang's third sound trope of screams and scream-like noises takes on more complicated patterns of signification than the use of either thunder or echoes. For example, screams are some of the first sounds to intrude upon the Don's inner sanctum at the beginning of the film when the arrival of Johnny Fontaine is heralded by the cries of his adoring fans. Initially introduced as benign, this sets up the use of screams for horrific purposes after Woltz discovers the head of his priceless steed Khartoum under his bed linens. Woltz's screams of terror are also heard in perspective, changing their reverberant characteristics when the camera cuts to the exterior of his estate, and overlapping a dissolve to a placid Don Corleone. Elsewhere, the crying of babies is used as synecdoche for the fear the adult characters feel. When Don Corleone is shot by Tataglia's men, Fredo looks on impassively as a barking dog and crying baby are heard offscreen. Shortly thereafter a loud crash is heard outside Sonny's house, which wakes his son who is heard crying from the back bedroom. When the Don returns to his house after leaving the hospital, another crying baby is heard offscreen, and when Carlo brutally beats Connie with a belt the film cuts from her screams to her mother holding Sonny's crying baby as she answers the telephone. In each instance the cries signal much more than just their sound. They telegraph the fears of the characters during each scene while also reinforcing the dominant theme of family within the film.

The scene most discussed in relation to its sound use is Michael's meeting with Solozzo and Captain McCluskey in an out-of-the-way Italian restaurant. Solozzo and McCluskey expect to meet with Michael, then considered a "civilian" rather than an active member of his father's organization, to discuss a truce. However, the audience knows that a gun has been planted in the restroom and we wonder whether Michael will be able to complete the task of assassinating his father's rivals. The audience also knows from a prior scene that Clemenza left the gun noisy to scare "any pain-in-the-ass innocent bystanders away." The scene represents Michael at his most taciturn, listening to Solozzo plead his case while also working through the logistics of the hit in his mind. Because this is difficult to express with just visuals, Walter Murch struck upon the use of a single sound effect to express Michael's inner turmoil. On his arrival at Louis's Restaurant, the sound of an overhead subway train is heard even though it is not seen. This fits with the realism of the period and helps to reinforce the restaurant's location in the Bronx, and it introduces the motif of the train, which Murch used to signal Michael's thought process.

The restaurant provides Solozzo with the privacy required for a clandestine meeting while also giving Michael the security of a public space; however, its acoustic tone sets up a tension that is palpable from the beginning of the scene. Not only are few diegetic sounds heard in the dining room, but certain effects are heightened in the soundscape. As Michael and Solozzo stare at each other across the table, a waiter uncorks a wine bottle. The sounds of the corkscrew laboring with the bottle and the inevitable pop of the cork reinforce the underlying tension, what Michael Ondaatje calls "a perverse celebration of a minor detail at a tense point."[18] As the scene continues, Solozzo speaks to Michael in Italian, yet Coppola refuses to subtitle the dialogue for an English-speaking audience. The resultant aphasic effect forces us to pay attention to the sounds around Michael rather than to the semantic meaning of the conversation. The film masterfully places the viewer in a mode of reduced listening as a second train is heard rumbling in the distance. As Murch commented on this moment, "That was very deliberately done, to make you pay attention to a tiny realistic sound and then have an overwhelming sound that you just have to interpret in a different way—all on a subconscious level."[19]

More than just interpreting the scene in a different way, it filters the scene through Michael's consciousness and we hear the sounds the way that Michael hears them, not specifically in terms of a realistic point of audition, but as a representation of Michael's attentiveness. When Michael goes to the bathroom to retrieve the gun, the sound of the overhead train is heard louder and more directly than before. From a logical standpoint this could be because the train line is at the back of the building—the sound is diminished when the film cuts back to the dining room—but the audience associates the sound with Michael going ahead with the plan. As the scene unfolds the camera remains on a pensive Michael while Solozzo continues speaking in unsubtitled Italian. Tension is built through a combination of a slow track in to Michael in medium close-up, Michael's darting eye movements, and the return of the overhead train sound. In this final instance, instead of the low frequency rumbles, the train sound contains high frequency scrapes and metallic groans that build to a climax when Michael kills Solozzo and McCluskey. Only after he drops the gun and leaves the restaurant does the score music kick in to reinforce the impact of the event.

The conceit of the scene is that sound effects were used to reinforce the period realism of the restaurant while also shifting audience perception to interpret the sounds as an analog for Michael's tortured thought process. As Judith Vogelsang explained: "The sound imagery is committed to the inevitable structure of the film. Because Michael has committed himself to the family, he must go through with the plan. He stands up and shoots them both as Clemenza instructed—twice in the head. The subway sounds from the scene, then, are the *integral, realistic* source for the use of the thunder and screech motif in *The Godfather*. The motifs in the film are not random or contrived. They flow naturally out of and into the structure of the work."[20] Already primed by the restaurant scene, the audience becomes attuned to the significance of other sound effects throughout the film. The reverberant drips of water in Bonasera's funeral parlor match the tiled interior of the room, but they also serve as an aural association to remind us of Sonny's bloody ending. When Michael, exiled in Sicily, hears about Sonny's death, his new bride Apollonia is playing with the car and honks the horn at an inopportune moment. The effect reinforces her childlike nature, and the impact of Sonny's death is soon redoubled after this momentary distraction when Apollonia is killed by a bomb planted in the same car.

Ironically, some of the most expressive uses of sound in *The Godfather* come from the absence of sounds: the absence of ambience at the beginning, the absence of dialogue during key passages, and the absence of music during major dramatic moments. A strikingly effective moment used to defer and reroute our generic expectations of sound occurs during the aforementioned meeting between Luca Brasi and Solozzo. After he has been instructed by Don Corleone to persuade the Tataglia family to take him into their confidence, the audience expects the hulking Brasi to have the upper hand. When, in a moment of abrupt violence, he is pinned to the bar by Solozzo's knife and strangled from behind, we suddenly realize that Solozzo and the Tataglias were one step ahead of Brasi and Don Corleone. "When Luca Brasi struggles in vain against the silk noose flung around his neck, Coppola's camera gazes at his protruding eyes like some mesmerized onlooker," noted Peter Cowie. "The absence of music, and the studied realism of the soundtrack (glasses crashing to the floor, Brasi's hoarse, animal groaning), reinforces the shock of the murder."[21] Cowie brings up two important elements of sound use in this one instance. Coppola constructed a strong sense of realism not only through the sets and props but also the sounds that accompany them. Luca's entry to the Tataglia bar is heard with his heavy shoes scuffing across the tile floor and echoing down the gilt art deco hallways. Also, the absence of music lets the audience shift from the sudden shock of the event to an awareness of its narrative significance.

Walter Murch expressed how the inclusion of music after a dramatic narrative event is used as a motif throughout the film:

> The general tendency in *The Godfather* is to play big scenes in silence and then to bring the music in afterwards. For instance, the killing of Carlo, Michael's brother-in-law, at the end of the film has no accompanying music. In a so-called normal film you would have the dramatic murder music, but we had only that sound of Carlo's feet squeaking on the windshield as he's being choked to death. Then his foot smashes the glass and you're left with the image of his foot sticking through the windshield and the sound of the gravel crunching as Michael walks back to the house. Then the music comes in.[22]

This repeated pattern lets the event sink in first so that the audience can work through the narrative implications before the music reinforces the thought process. To amend Murch's statement slightly, the scenes do not play out in silence; rather, they rely on sound effects and dialogue to convey meaning rather than using "murder music" to control audience reactions. In this way it allows the film to keep the audience in suspense by not signaling when violence will occur. From a narrative perspective, the violence feels sudden and random rather than the programmatic way the gangster genre used music to build dramatic tension.

In contrast, after developing the motif of nondiegetic music occurring only after a dramatic event, Coppola reversed the formula for the extended baptism sequence at the end of the film. Here, the diegetic organ music begins as a marker of the baptism of Michael Francis Rizzi, Connie and Carlo's son, but rapidly becomes a harbinger of the violence to follow. What starts as nondescript liturgical music played on a church organ slowly transforms into Bach's Passacaglia and Fugue in C minor (BWV 582) as the images cut from the rituals of the baptism to Michael's gunmen preparing to murder the heads of the five families. The multiple views of his capos are intercut with the baptism, and the organ and Latin recitation carry over each of the cuts. Initially the individual vignettes seem benign—Willi Cicci getting a shave, Al Neri trying on a suit, Clemenza leaving his house with a large package—but a shift in the music from major to minor when Rocco Lampone is seen cleaning his machine gun signals to the audience that things are not as they seem. The rumbling bass ostinato from the Passacaglia begins the moment when Neri empties a paper bag onto his bed, revealing a revolver and policeman's badge. Bach's piece provides an underlying musical tension that contradicts Michael's abjuration of evil during the ceremony, and the cries of Michael Francis resound to mark the generational transition of power enacted by the gunmen. Michael's three renunciations are followed by excessively bloody acts of violence that purge his enemies in a single campaign, each event heard directly with the organ music carrying over. Because the music overlaps the cuts, it connects the actions of his men back to Michael and his heretical vows. At the moment when Michael becomes godfather to Connie's son we realize he has ascended to the station of Don, the Godfather of the Corleone family.

The Godfather Part II (1974)

Dramatically more complex in the way that it interweaves parallel stories of Vito Corleone's emigration to America and rise to success with Michael's actions as Don, *The Godfather Part II* also relied heavily on sound to aid the storytelling process. The sound tropes used in *The Godfather* returned in its sequel where they were expanded upon and modified. More than the original, the sequel amplified the evocation of space through reverberation. Murch discussed the origins of the trope with Michael Ondaatje, recalling, "I was already fascinated with the ability to shift perspective in sound, particularly in something like the wedding scene in *The Godfather*—the shift from the noise of the wedding outside, to hear different perspectives."[23] Providing valuable information on Vito's backstory, *The Godfather Part II* quickly moves to Sicily in 1901 where Vito's father was killed in an escalating familial vendetta. His funeral is introduced by the sound of a brass band echoing off the bluffs as the procession makes its way through a dry riverbed. Distant gunfire and reverberating reports erupt to be replaced by the wails of an unseen woman announcing that the rival gang has killed Vito's brother Paolo. That night the voices of the town criers echo through the town of Corleone, urging the citizens to turn over young Vito to the local patriarch, Don Ciccio. After being smuggled out of town, Vito arrives in the United States and his first experience is waiting in the interminably noisy and reverberant processing center at Ellis Island. As in the first film, reverberations are used to foreshadow dramatic events or to remind viewers of the potential eruption of violence.

Several scenes from the first film are paralleled in the sequel as are their commensurate sound devices. After Vito's arrival at Ellis Island the film transitions to Lake Tahoe and the first communion of Michael's son, Anthony Vito. As in the wedding that opened *The Godfather*, the sound of Anthony's communion band changes according to the position of the camera to create a verisimilitude of space and time. When Michael sends his bodyguard to kill Hyman Roth in a Havana hospital, the reverberant hallways and distant moaning of patients set up a mood that closely matches Michael's visit to his father's hospital room in *The Godfather*. At the end of the sequel, Michael shuts Kay out of his life with the wordless gesture of a door closing and then proceeds to consolidate his power by killing his rivals and his brother Fredo. This penultimate sequence, however, reflects one of the major differences in *The Godfather Part II*. Instead of diegetically sourced music, Nino Rota's score music is used to carry over to the intercut scenes. In fact, the sequel relies on the use of score music to presage events and signal emotional responses rather than to serve as after-the-fact commentary. Curiously, by increasing the narrative complexity of the second film, Coppola retreated from some of his progressive experiments in *The Godfather* to provide a more familiar approach to sound and music use.

Instead of the metaphoric use of sound favored in *The Godfather*, the sequel occasionally resorts to some common clichés. After Anthony's communion party, Michael returns to his bedroom where Kay is already in bed. An ominous clap of thunder is heard, and Kay cues audience anxiety when she queries, "Why are the drapes open?" moments before a fusillade shatters the windows. Shortly thereafter the bodies of the gunmen are found when Fredo's wife, Deanna, is heard screaming outside their cottage. During the film's middle section, set in Cuba on the eve of the overthrow of President Batista, the sound of sirens is regularly heard in the distance, moving ever closer to the city center. And upon Michael's return to Las Vegas, the revolutionary uproar of Havana is contrasted with the desiccated desert wind and the hermetic quiet of Michael's office. Most tellingly, Rota's music is used liberally throughout the film to underscore transitions and to support narrative events, a technique at odds with the reactive use of music in the first film.

Despite what can be heard as a slight retreat from the innovative sound practices in *The Godfather*, *The Godfather Part II* continues an emphasis on the significance of the voice in performance. The casting of Robert De Niro as a younger version of Vito Corleone relied less on his physical resemblance to Marlon Brando than on his ability to internalize Vito's speech patterns and understated delivery. As Chion observed, "Robert De Niro took up the role of the young Corleone and was forced to invent a voice—speaking almost entirely in Sicilian—that was compatible with Brando's voice in the first *Godfather*."[24] More than anything, this vocal masquerade reinforced the link between De Niro's and Brando's renderings of Vito Corleone while also cueing the audience to note parallel changes in Michael's voice and delivery. As the young Vito discovers his true voice—his own mother claimed him to be "dumb-witted" because he never spoke—so too does Michael. In *The Godfather Part II*, Michael's laconic penchant is expanded and his dialogue tapers off to a trickle by the end of the film. Just as the film's visual style moves further into murky darkness, so too does Michael's reticence.

The audience therefore understands the special significance carried in the few lines that Michael does speak. After returning from the failed Havana venture, Michael hears that Kay has lost the baby she was carrying, and instead of asking after her health, Michael's only response is, "Was it a boy?" Not only does this echo the misogynist sentiments of the first film ("I hope that their first child be a masculine child," "Women and children can be careless, but not men"), but it also shows the level to which Michael and Kay's relationship has deteriorated. Moreover, Michael, unlike his father, is unable to keep his anger in check and his progressive unraveling is signaled through a series of vocal outbursts. His explosive argument with Kay in a hotel room bleeds through the walls and is first heard from the perspective of Anthony and his sister Mary in the hallway. By the end of the film the audience comprehends the depth of Michael's compulsions when he only needs to say Connie's name as a warning to put his wayward sister back in line. Other than this single utterance, the consolidation of Michael's power in the penultimate sequence takes place without him saying a word.

One of the reasons why the sound strategies in *The Godfather Part II* were not as experimental as in the first film was because Walter Murch was concurrently editing and mixing another film for Coppola. Filmed in early 1973, *The Conversation* (1974) finished initial production at the same time that Coppola embarked on pre-production for *The Godfather Part II*. Because of the pressure placed on the director by Paramount Pictures, Coppola left *The Conversation* in the hands of editor Richard Chew with Murch supervising its post-production. As Peter Cowie pointed out, "A formidable responsibility lay with Walter Murch during this period. Not only did he have to ensure that *The Conversation* would emerge as planned, but also supervise the sound design of *The Godfather Part II*."[25] Although Murch can be considered Coppola's collaborator on *The Conversation*, he is better thought of as a cocreator. Because the story centers on a surveillance expert who becomes embroiled in a conspiracy discovered through an audio recording, the film relies on the audience hearing the world the same way that the central character Harry hears it. Drawing on his experience with *The Rain People* and *The Godfather*, Murch created a hermetic world where the audience experiences events through Harry's subjectivity.

The Conversation (1974)

The primary concern addressed in *The Conversation* is how to render hearing in a primarily visual medium. This is obviously not a simple question to answer, and the depiction of Harry Caul (Gene Hackman) recording and reconstructing an illicit conversation for an unknown employer places him at the center of a crucial matrix of ideologies. First, his career resonates with what were then recent events of the Watergate bugging and the revelation of the Nixon White House tapes, the full extent of which was unknown when Coppola made the film. Although its basic premise was formulated in 1967 and filming completed in early 1973, the film takes on extra significance due to its release mere months before Nixon's resignation. Second, Harry's inability to relate to the world around him is indicative of the sense of disconnection present in American life during the 1970s. Third, as a way to isolate him from the world surrounding him, Harry relies on technology as a protective shield. Finally, as Robert Kolker wrote, "[Harry] is like a filmmaker, putting together bits and pieces to make a whole."[26] Not only does he stand as narrative surrogate for Coppola the filmmaker, but his work in reconstructing a complete conversation out of pieces of recorded dialogue emulates the act of sound editing and mixing for films. As Kolker averred, however, "What he puts together is the wrong movie."[27]

The Conversation functions as an exploration of Harry's life rather than the lives of those he is spying on. The audience is introduced to Harry in a unique way. A slow long-take zoom from a rooftop reveals a number of people wandering around San Francisco's Union Square until the camera finally settles on Harry moving through the crowd. Simultaneous with this action is the construction of an audio-zoom that increases the sounds in volume to match the changing shot scale. On the audio track are several disturbing sound effects, sounds that we later discover to be the digital interference caused by the high-powered and aptly named shotgun microphones. As quickly as the audience is introduced to Harry, the focus of visual attention shifts to two characters, played by Cindy Williams and Frederic Forrest, as the recording picks up their incomplete snippets of dialogue. This mismatch between sound and image not only serves the surveillance narrative but also foregrounds the constructed nature of the sound track itself. The film relies on this deconstructive gesture as a way to open up a sense of doubt in relation to the recording technology and Harry's perception. The sound mix is heightened to emphasize what things would sound like to someone who spends his life listening in on other people's conversations: everything is louder, more present than normal, and offscreen sounds are suppressed. Murch mixed the film's optical track for theaters with the loudest passages at a maximum modulation while refusing to use compression or expansion on the rest of the sound track. The result is a film that sounds unusually quiet, requiring the audience to listen attentively to the sound track until certain passages where the volume is startling. Murch expressed the effect by saying that "the most successful sounds seem not only to alter what the audience sees, but to go further and trigger a kind of *conceptual resonance* between the image and the sound: the sound makes us see the image differently, and then this new image makes us hear the sound differently."[28]

This "conceptual resonance" was the quality that many filmmakers of the 1970s sought, the ability to unite the formal aspects of the film with the narrative. Coppola's creative use of sound allowed him to connect the audience with Harry's perceptual state. When Harry is recording his material, content is irrelevant. Instead he wants to provide his employer with what he calls a "big, fat

FIGURE 11. Harry (Gene Hackman) listens too attentively as he reconstructs clandestine recordings in *The Conversation* (Francis Ford Coppola, 1974).

recording." All that concerns him is restoring the voices to a level of intelligibility, with no thought given to the actual words being said. His role here emulates that of the dialogue editor, whose job is to select the best tracks and takes, and the re-recording mixer, who mixes all the tracks so that every word of dialogue can be heard. Harry works like an artisan while reassembling the conversation from the extant recordings, but his attentiveness to the sound of the conversation prevents him, and the audience, from clearly interpreting the meaning behind the words (see figure 11). While Harry's job is to ensure the absolute intelligibility of the conversation, Coppola does not allow the audience easy access to its meaning. Even though a critical line—"He'd *kill* us if he had the chance"—is revealed by cleaning up the recording, its significance is not fixed.

As a way of adding a conceptual depth to the picture, Coppola and Murch played with the notion of intelligibility in cinema by making it difficult for the audience to hear all the lines in the film. Scenes take place in Harry's cavernous workshop where a plurality of voices and reverberations make comprehension difficult. Instead of having the characters each individually wired for sound or looped in post-production, Nat Boxer's spatially oriented miking strategies keep the audience aligned with Harry's perspective. The audience is deliberately not allowed to hear all the lines of dialogue, thereby emphasizing Harry's inability to communicate. Restricted to Harry's acoustic and visual perspective, the audience is reliant upon the accuracy of his perception while questioning the ability of the recording apparatus to record the truth behind the words. The central concern of Coppola's film is not necessarily what was actually said, but what Harry thought was said. It is his perception and its inherent fallibility that Coppola emphasized more than the murder plot exposed in the reconstructed recording.

This is done by restricting the narration to Harry's perspective, letting the audience believe that the technology revealed the truth behind the recording and then undermining that interpretation by exposing Harry's inability to interpret the words being spoken. Convinced that he has delivered evidence of his employer's wife's infidelity, Harry tries to intervene by bugging the hotel room where he fears that she will be murdered. However, when he hears his recording of the conversation being played and an ensuing melee, he is rendered incapable of action, paralyzed by his overwhelming guilt. Only in the penultimate scene does he realize that the tape was not about the couple fearing for their own safety, but rather their plans to kill her husband. This is discovered when Harry, and the audience, hears the line one last time in its true context: "He'd kill *us* if he had the chance."

Coppola and Murch understood the possibility of using film sound as a way to expand the story and to engage the audience on a higher level than through a simple correlation between sound and image. The recording of the conversation in the park was both a tool for advancing the narrative, with the materiality of sound forming the heart of the story, and a device that enunciates the constructed nature of cinema. Technology itself becomes an active agent, one that carries a simultaneous promise of the betterment of daily life and a threat of insidious ubiquity. However, it is not the technology alone that represents power and security in Coppola's film; it is those who have the ability to interpret its message. At the film's end, Harry is left alone in his destroyed apartment, torn to shreds in a vain search for a planted microphone. His protective shield of technology, turned against him, has become a prison of what David Denby called "stolen privacy."[29]

The use of sound in *The Conversation* was possible because of the freedom of the sound mixer to work closely with the director and to marshal the sound track to the service of the narrative. But this period of opening for cinema sound was short-lived in its potential. Concurrent with films like *Star Wars* in 1977, Dolby Stereo introduced new rules of film sound recording and mixing that effectively served to cover over the gap created by prior sound experiments. By separating out dialogue mixing and elevating it to the top of the post-production sound hierarchy, Dolby Stereo was a retreat from the creative construction of the sound track to a single strategy for mixing. The demands of the system meant that the practice of post-production sound changed from an act of artistic creation, often guided by one individual or a closely knit team, back to a regimented, hierarchized system of labor. Coincident with the introduction of a standard of post-production sound mixing is a marked change in the way sound is utilized in narrative construction. Narrative emphasis tends to be placed in a single acoustic register, favoring only one component of the final sound track, in lieu of the film's sound containing a valuable conceptual resonance. Audiences were interpellated through the familiar codes of classical cinematic representation into which Dolby Stereo had neatly assimilated itself. The result is a presumed continuity between the codes of classical Hollywood cinema and the films of the 1980s, thereby eliding and effacing the creative cinematic experiments of the 1970s. It is through films like *The Conversation*, *Taxi Driver*, *Nashville*, and especially Coppola's *Apocalypse Now* that the continuity of classical Hollywood form and post–Dolby Stereo mixing practices is complicated and crucial alternative models of representation are explored.

6

Robert Altman's Collaborative Sound Work

> The filmmakers that influenced me the most . . . I don't know their names. Because I would go see a film and hate it, and I'd say I gotta remember never to do anything like that again.
>
> –Robert Altman

During the first half of the 1970s, Robert Altman's career was marked by rethinking sound recording practices and organizing his production team as an egalitarian endeavor rather than along the lines of the hierarchical construction of classical Hollywood sound recording and mixing. More than just the oft-stated use of overlapping dialogue, Altman set about training his collaborators and his audience to see and hear the world differently. He broke from the conventions of dialogue intelligibility that dominated the classical Hollywood cinema and developed several strategies in sonic storytelling. First, he downplayed the significance of individual lines and asked audiences to listen to multiple speaking voices to sort out the narrative significance for themselves. Second, he developed a taste for the use of direct sound rather than looping dialogue in post-production. Third, he emphasized offscreen sounds and often structured a dialectical relationship between the sounds offscreen and their effect on onscreen characters. Fourth, he introduced the practice of having meta-diegetic sounds—voices and music that exist within the diegesis but comment on the story—as a reflexive form of authorial commentary. This tension between the sound and images in his films is the main marker of his sound practice during the period from *M*A*S*H* (1970) through *Nashville* (1975). Accordingly, his complexly layered sound tracks are indicative of an attempt to restructure the way in which audiences watch and hear motion pictures.

Brewster McCloud (1971)

Ever since its release, *Brewster McCloud* has split critical responses, with some declaring it an unmitigated flop, and others, Altman included, looking back on it fondly. For the director this may have to do with the extreme latitude afforded him by MGM after his surprise commercial success with *M*A*S*H*. Despite his claim that "it was my boldest work, by far the most ambitious,"[1] *Brewster McCloud* stands as a transitional film when it comes to Altman's visual and acoustic aesthetics. Like *M*A*S*H*, its narrative is fragmented and the picaresque emphasis on the titular character often serves as an excuse to string together a variety of comic sketches and non sequitur scenes. Yet, in its own way, the film's story about a boy who longs to fly lets Altman fully explore his sound gags and metadiegetic narration, albeit not without ruffling a few feathers.

In a practice that carries on throughout many subsequent Altman films, *Brewster McCloud* tweaks a sacrosanct Hollywood tradition by interrupting the opening studio credits. Instead of the legendary roar of Leo the Lion over Metro-Goldwyn-Mayer's "Ars gratia artis" motto, the audience hears "I forgot the opening line" as the film reveals René Auberjonois's bird lecturer in a classroom. Because the spatial and temporal relationship of these lectures remains unknown throughout the film, their appearance functions as a form of metadiegetic commentary on the main narrative. Initially seen standing in front of a blackboard, the lecturer continues to be heard over several passages in the film and his narration offers a comic contrast to the images. When he starts to discuss the courtship rituals of birds the film cuts to the funeral of an undercover police officer. The voiceover, describing the male bird's mating dance, draws our attention to the flirtatious exchanges between the officer's widow and his former partner. Although the narration redirects our awareness and undercuts the seriousness of the images, in *Brewster McCloud* it is purely for comic effect and far from a dialectical or critical strategy.

Indeed, much of the film is based around a gag logic that uses sound to underpin the comedy. After the bird lecturer prologue, the film begins in the Astrodome where an African American marching band plays "The Star-Spangled Banner" while Daphne Heap (Margaret Hamilton) sings at the microphone. Realizing the song is in the wrong key, Hamilton's character, suspiciously similar to her Miss Gulch/Wicked Witch of the West from *The Wizard of Oz* (Victor Fleming, 1939), reproaches the band and orders them to return to the beginning. The music starts anew as the title credits appear a second time. Altman not only wants his audience to recognize the fabricated nature of the film, reminding us of the technical elements of its construction, but he also asks us to be aware of the significance of the music we hear throughout. In this case, the national anthem is interrupted when Brewster (Bud Cort), who lives in the

Astrodome's bomb shelter, plays a recording over the PA system of Merry Clayton's "Lift Every Voice and Sing"—also known as "The Black National Anthem"—to drown out Heap's singing. Another musical gag attached to Hamilton's character occurs after she is "murdered" while tending her prize pigeons. Seen in flashback *after* a radio news broadcast relays the general details of her death, Heap is killed when a crow removes a support pin from her wrought-iron aviary and it collapses on top of her. As the radio announcement continues, describing what she was wearing at the time of her death, the camera slowly zooms into her ruby slippers while the sound track plays an instrumental version of "Somewhere Over the Rainbow," thereby solidifying her connection to *The Wizard of Oz* and underlining the gag.

While both the interpolated bird lecturer and the radio news updates continue throughout the film, the organizational structure behind them is relatively unclear. This likely has to do with Altman still developing his approach to sound and image relations in his films, and unlike *M*A*S*H* there seems to be little attempt to make overlapping dialogue part of the aesthetic. As Robert Kolker pointed out: "Two sequences within this film—one in a police laboratory, the other a police investigation of a murder on the street—are constructed with large numbers of people talking all at once and at cross purposes, bad jokes weaving in and out of the conversations, no one element taking precedence over the others. These sequences tend to be isolated, for Altman is working out other problems of narrative structure."[2] Kolker's astute observation about narrative structure can also be applied to the sound techniques in the film, several of which seem to be consonant with Altman's prior work and others that seem to be discordant with his emerging aesthetics.

One practice that gets reinforced is an emphasis on location sound and an attempt to balance the foreground dialogue between the main characters with a strong awareness of background conversations. As in *M*A*S*H*, this was accomplished by using several microphones at the same time to record a variety of sounds during production. Journalist C. Kirk McClelland was granted exclusive access to the set during the production of the film, and his book on the making of *Brewster McCloud* provides a unique snapshot of the production process: "The sound men on this film are Doyle Hodges and Bob Kaplan. Doyle is the sound mixer and Kaplan handles the boom-mike, which is the primary mike in almost every shot. I introduce myself to Doyle and ask him a few questions about his equipment. He tells me that for the last scene he used three mikes. Kaplan placed two mikes around the room and then held the boom-mike over Wright and Brewster as they traveled."[3] The presence of the Houston-based Hodges and Kaplan characterizes the standard practice of hiring most of the production crew locally and bringing in only the camera department. While their use of three microphones to cover the background sounds as well as the primary dialogue is innovative for the time, the production mix clearly favors the foreground conversations of the central characters over background subconversations. Regardless of the emphasis placed on its meta-diegetic commentaries, *Brewster McCloud* stands as a curious hybrid of a film where Altman's sense of narrative and audiovisual aesthetics is still being worked out. Yet it also serves as a strong example of Altman developing new sound practices and training his crew in his evolving methods.

McCabe and Mrs. Miller (1972)

After the experimentation of *Brewster McCloud, McCabe and Mrs. Miller* marks Altman's first deliberate attempt at developing new audience viewing and listening patterns. Instead of presenting the audience with a clear plot and easily identifiable character motivation, Altman puts viewers in the position of having to piece together fragments of narrative information in order to discern the entire story. This is done through the practice of overhearing rather than having dialogue directly presented to the audience. Set in the Pacific Northwest around the turn of the twentieth century, the film documents the growth of the small mining town of Presbyterian Church by following newcomer and entrepreneur John McCabe (Warren Beatty). The period flavor of the town, which was actually built during the shooting of the film, is seen in the sepia tint of Vilmos Zsigmond's cinematography and matched by the overlapping of voices and sounds in each scene.

Returning to the strategy from *M*A*S*H* and *Brewster McCloud*, Altman decided to record all the voices on location and let the lines of dialogue overlap each other almost to the point of unintelligibility. Vancouver-based sound mixer John W. Gussele was recruited to record the voices on location, and instead of relying on lavalier mics he used shotgun microphones at a distance to preserve the spatial characteristics of the voices. This resulted in a sound that was realistic to the wood-paneled sets yet introduced a level of reverberation that made it difficult to hear individual lines of dialogue clearly. Even though this can be seen as a deficit, it was a technique that Altman cultivated as a signature sound for his films, starting with his earlier collaboration with Gussele on *That Cold Day in the Park* (1969).

In addition to the use of strong spatial cues, Gussele also recorded several background conversations that did not directly advance the main plot. These conversations were just as important to the development of characters as costume or makeup, and the fragments of conversations heard from the numerous characters in the community helped to explain who John McCabe was and his motivations. Because many spoken lines of dialogue were overheard it was almost impossible to attribute these subconversations to specific characters.

FIGURE 12. Not only are the spatial signatures of the locations emphasized in *McCabe and Mrs. Miller* (Robert Altman, 1972), but so too are the fragments of overlapping dialogue that speculate about John McCabe's (Warren Beatty) backstory.

Instead, the overheard lines reinforce the effect of rumors about McCabe, "the man who shot Bill Roundtree," which add up to become truths about his character (see figure 12). René Auberjonois, who played Sheehan in the film, noted that in Altman's sound practice "you don't need to hear everything people are saying to know the world they're living in."[4] The combination of overheard dialogue and the spatialization of voices worked to tease out the narrative details through the acoustic fragments.

Reflecting on the process, Vilmos Zsigmond expressed his initial reservations with Altman's approach to dialogue:

> The sound track was very, very courageous because he deliberately made the sound so too many people are talking at the same time. I even questioned it myself. I said, "Robert, the sound mix, I can't understand what they're saying." He said, "But have you ever been in a bar where there's so much noise, so many people arguing, do you hear everything that they say three tables away? Well, that's what I try to do. That's exactly what I want to have, the feeling of reality. Not that clear, perfect, beautiful sound recorded on a soundstage."[5]

Although he did not articulate his strategy in interviews, Altman did explain the rationale for the background conversations in *McCabe and Mrs. Miller*. "The first forty minutes of the film is really nothing but atmosphere, all of which is planted, like collateral, for the audience to draw upon for the last hour or so."[6] What initially seems like inconsequential information takes on new meaning with the arrival of three gunmen hired to take away McCabe's business interests by force. After hearing rumors about McCabe's past as a gunslinger—which he never acknowledges or bothers to deny—McCabe is finally challenged by the lead gunman, Butler (Hugh Millais), who accuses him of killing his "best friend's best friend," Bill Roundtree. Confronted with the very myth that he used to leverage his position in the town, McCabe backpedals before retreating from the saloon, leading Butler to declare, "That man never killed anybody."

The remainder of the film finds McCabe defending his reputation and his life from the encroaching threat of Butler's men. After an hour and a half of overabundant dialogue, the last act of the film plays out over the sounds of a snowy morning in wordless ambient passages. "The lack of dialogue [in the last sequence] was just that I don't know what people are going to say in this sort of situation," noted Altman: "What we did do that was sort of adventurous was not to use any extraneous sounds. The last 23 minutes of the picture are virtually silent. That's a very dangerous thing to do because when it's quiet you're not covering up your audience noise. We just gambled we could hold 'em with the suspense."[7] The effect is quite powerful as McCabe valiantly fights for his life while the rest of the town is engaged extinguishing a fire at their namesake church. The counterpoint between the community banding together and McCabe fighting alone is reinforced by the lack of dialogue and creates a strong parallel with McCabe's entry into the town. Yet this time, instead of riding off on horseback, the same way he entered the story, McCabe lies in a bank of snow, mortally wounded, unseen and unheard by anyone as he becomes obscured by the drifting snow.

As part of Altman's larger sound strategy, the lack of directly relevant narrative dialogue early in the film is made up for by the inclusion of nondiegetic musical commentary. As Paul Arthur explained, the music functions as "commentary in the form of a Leonard Cohen song (there is no instrumental sound track, although diegetic music is performed in several scenes) that connects McCabe with Christian martyrdom, anticipates the arrival of McCabe's love interest, and cues the dominant theme of business or 'deal-making.' Cohen is no Fordian Sons of the Pioneers and his four songs pierce the dramatic action with a sweetly wistful morbidity."[8] Though Arthur's musical accounting is off by one, the three Leonard Cohen songs in the film continue the trope of metadiegetic commentary in Altman's works. Unlike other films that used popular music tracks as nondiegetic scores to supplement character development (*The Graduate* [Mike Nichols, 1968], *Easy Rider* [Dennis Hopper, 1969]) or as diegetic markers of their contemporaneity (*Petulia* [Richard Lester, 1968], *Two-Lane Blacktop* [Monte Hellman, 1971]), Cohen's music in *McCabe and Mrs. Miller* has an uncanny frisson with the characters and the story being told.

As he wanders his way into town accompanied by "The Stranger Song," most of McCabe's backstory is gleaned from Cohen's lyrics. To Altman's credit, he uses the song as a support element rather than the main framework for advancing the story by cutting up the song's structure and reassembling the lyrical passages to make apposite connections. For example, the first three verses of the song are heard as McCabe approaches the town, and passages like "It's true that all the men you knew were dealers / Who said they were through with dealing / Every time you gave them shelter" and "Like any dealer he was watching for the card / That is so high and wild / He'll never need to deal another" describe him as a card sharp who seeks to ply his trade in Presbyterian Church. As he rides through town, we hear an artfully re-recorded guitar passage inserted into the original song to allow McCabe's mumbles to himself to come through on the sound track. As the lyrics continue, the fourth and fifth verses of the song reveal McCabe's goal "to trade the game he plays for shelter," as he dismounts and heads to Sheehan's Saloon to set up his poker game.

Altman continues this process by using Cohen's "Sisters of Mercy" to introduce the prostitutes brought up from Bear Paw to work for McCabe and "Winter Song" to underscore the complex relationship between McCabe and Constance Miller (Julie Christie). Because these songs are entirely atemporal and nondiegetic, they provide important narrative and character information from outside the diegesis. Cohen's songs function as both score music, underpinning the emotional resonances with the narrative, and as commentary on the main action of the film. In this way they can also be called meta-diegetic, yet unlike the songs over the squawk box in *M*A*S*H* or the radio announcements in *Brewster McCloud*, they are not diegetically sourced. This process of meta-diegetic commentary is a main marker of Altman's films throughout the 1970s, though it will take several films before it achieves its final dialectical form. In the case of *Images*, Altman's sound techniques are almost entirely based around rendering the mental state of the main character through sounds, voices, and music.

Images (1972)

Images is arguably the film that has received the least amount of critical writing among Altman's films from the early 1970s. Like *That Cold Day in the Park*, it tells the story of an emotionally disturbed woman and ends with her committing an act of violence. After the plaudits received for *McCabe and Mrs. Miller*, *Images* received a critical drubbing. Even though it can be argued that the film's overall narrative arc reinforces claims about Altman's proclivity toward sexism in his films, a closer look reveals a nuanced attempt to convey the interior emotions of the film's central character, Cathryn (Susannah York), by rendering her perception of the world onscreen and through sound. To achieve this, Altman relied on an unusual blend of visual and acoustic distortions to represent Cathryn's tenuous grasp on reality.

Made for Helmdale Studio in Britain, the film contains a number of practices new to Altman's work. First, as was the norm in British filmmaking during the late 1960s and early 1970s, almost all the dialogue was replaced in post-production, leaving a strange airless quality around the words that fits with the oneiric story being told. After the vibrantly real sounding locations of *McCabe and Mrs. Miller*, the conspicuously dry ambiences in *Images* work to direct the audience's attention to Cathryn's mental state rather than reinforcing diegetic verisimilitude. The absence of spatial signatures also draws attention to the heightened presence of selected sound effects, especially the acoustic motif of wind chimes that recurs throughout the film. Specifically, the sound of wind chimes becomes attached to the prismatic reflection of crystals in the film, grafting the audio motif onto a visual phenomenon. Though we are used to sounds being attached to non-sound producing objects for emotional effect through synchresis, this form of audiovisual schizophrenia hints at the larger sound strategies at work in the film.

Images stands apart from Altman's early works because all the music used in the film is strictly nondiegetic. There is no visualized source for the music, yet it tends to change and respond to Cathryn's mental state through a split between two distinctly different musical forms. Altman explained his working process as follows:

> I let [John Williams] read the script and suggested to him that he record his music and then I'd shoot the film and use his score. We didn't completely do that, but it's a method I've employed since. I'm not much interested in music that just goes along with the action. John came up with Stomu Yamash'ta, a percussionist who would do an act creating all sorts of sounds, like throwing a rock on to the strings of a piano. In his score, John would write descriptions of sounds that [Yamash'ta] would then perform.[9]

The film regularly interpolates Yamash'ta's atonal noises into Williams's tonal scores to represent Cathryn's growing inability to distinguish reality from fantasy. While it is a bold method for expressing her shifting mental state, the musical oscillations quickly border on cliché. This relatively simplistic idea of music versus "noise" maps onto Cathryn's progressive drift from sanity to psychosis, yet, as Judith Kass pointed out, "instead of commenting on what is happening, the noises interfere with one's concentration. *Images* is a film that should be seen stoned, or with every pore open and alert to catch each nuance. The sounds get in the way of the nuances."[10] Williams's use of Yamash'ta's noise compositions functions as a form of commentary on Cathryn's mental state, though it does little more than anticipate her delusional visions. In many ways *Images* is a return to an acoustic style attempted in *That Cold Day in the Park* and subsequently discarded. Even though Altman would return to a subjective use of sound in *3 Women* (1977), he would first develop a form of dialogue recording and reproduction that would make possible a simultaneous subjective and objective sense of sound in *The Long Goodbye.*

The Long Goodbye (1973)

Unlike the live location recording done on *McCabe* and *Brewster McCloud* or the looped dialogue of *Images*, *The Long Goodbye* strikes a curious balance between the two. The story, adapted from Raymond Chandler's 1953 novel, finds detective Philip Marlowe (Elliott Gould) transported to the 1970s. Instead of the book's first-person narration, Marlowe's internal thoughts are conveyed through a monologue spoken under his breath and often heard by no one except the audience. For this technique Altman combined location and looped dialogue to best accommodate the needs of the narrative. Whereas many of the main lines of dialogue are marked by their spatial characteristics, often heard reflecting the sound of the locations, nearly all of Marlowe's commentary monologue is heard directly and without reverberation. In this way Altman naturalized the traditional voiceover commentary of the detective genre and rooted it in Marlowe's character and the contemporary Los Angeles setting.

In part this is due to Altman's nontraditional take on the detective genre and the near complete lack of ratiocination on Marlowe's part. Instead of finding clues and piecing together evidence, Marlowe discovers the truth through movement and accident. He does not uncover the web of deception that surrounds him so much as he stumbles into it. The problem is that the clues are hiding in plain sight and Marlowe is unable to discern their patterns until after it is too late. Just as Altman packs the visual field with details, he makes the sound track bubble with what Donald Lyons refers to as a "polyphonic hubbub" of voices, which the viewers, like Marlowe, must sift through.[11]

Even though Altman was using established technologies and techniques to record sound during production, the way that he deployed dialogue and sound effects was unique for the time. As Judith Kass explained, the sound strategies in *The Long Goodbye* presage the multiple speaking voices and polyphony of Altman's later films: "The sound for *The Long Goodbye* was handled by John V. Speak in a conventional mike-and-boom manner, but it sounds as though Lion's Gate 8-Track Sound was used. . . . It isn't 'pure'; there's room noise, overlapping dialogue, and at times a cacophony of sounds, at other times an almost artificial quiet, as when Marlowe arrives at night and searches Dr. Verringer's rest area for Wade."[12]

The Long Goodbye marks an important moment in the evolution of Altman's sound practices in which the director was able to achieve his desired effects within the framework of the existing system of production. Yet as Kass foreshadowed, the limitations of the contemporary production capabilities pushed Altman to further refine his sound techniques in 1974 and 1975 by using new multitrack technologies. In the case of *The Long Goodbye*, the film is marked by his desire for the audience to hear a variety of competing conversations as well as Marlowe's reactions to them.

While the recording and deployment of dialogue in the film is an important step in the evolution of Altman's sound practices, the most commonly noted sound device in *The Long Goodbye* is its ubiquitous theme song. Written by John Williams and Johnny Mercer, "The Long Goodbye" appears in no fewer than ten musical variations across the film. As Marlowe leaves his apartment at the beginning of the film, the Dave Grusin Trio's version is heard drifting over from his neighbor's flat. When the film cuts away to Terry Lennox's (Jim Bouton) car exiting the Malibu Colony, Jack Sheldon's vocal version of the theme is playing on the radio. Cut back to Marlowe in his anachronistic 1948 Lincoln Continental listening to vocalist Clyde King's wispy piano ballad version. Cut in mid-verse to Marlowe entering a supermarket to hear, in sync with the prior versions, a Muzak recording of the theme. "I thought that was another brave idea. To make that song the whole score," noted fellow director and costar Mark Rydell. "A million different ways it was played, including with a Mexican funeral band. Doorbells, Muzak, everywhere."[13] This bold technique is credited to Altman by composer Williams: "He said, 'Wouldn't it be great if there was one song, this omnipresent piece, played in all these different ways?' We would go into a dentist's office or an elevator and there would be this ubiquitous and irritating music playing. It was threaded through, kind of like an unconscious wallpapering technique. I think it's completely unique. I don't think anyone has tried it quite the same way before or since."[14]

At the same time that Altman's musical conceit draws attention to its own presence, it also marks a progression toward sourcing all the music in his subsequent films. "I've always said at the beginning of conceiving a film, 'I'd love the music to be indigenous, so that there's not going to be any violins that you can't see, that it won't come from nowhere,'" noted the director. "I've never completely achieved that, though in *The Long Goodbye* the music becomes a character in itself."[15] His films from the 1970s show a growing resistance to nondiegetic score music as a device for manipulating audience emotions. As Altman explained: "The reason why we have to have music in films is to put a cocoon around us, so the audience doesn't become conscious of other people or embarrassed by being there. The music is a kind of tunnel to help keep your focus. . . . One day I'll do a film without any music."[16] Although that day never came, his subsequent films revitalized and remobilized the way that speech and music are used in narrative films.

Thieves Like Us (1973)

Set in the 1930s during the rise of network radio stations, *Thieves Like Us* examines the way radio shaped American life through the effect the medium has on the characters. Radio broadcasts are ubiquitous throughout the film and function as both meta-diegetic commentary and a surrogate for the musical score. As Judith Kass explained, "Altman uses old radio programs as a thematic counterpoint and a sort of second narrative for his film (there is no music score)."[17] This second narrative runs in contrast to the get-rich-quick schemes of the main characters, and its use interrogates how radio cultivated a new form of American culture. Like the characters in his film, Altman was weaned on the messages beamed over the airwaves: "At that time in the 1930s that's all I did as a kid, listen to the radio for two hours when I got home from school. Radio was everywhere, filled with commercials—it's what created the consumer society."[18] *Thieves Like Us* locates the roots of American consumerism in the Depression era, and the radio broadcasts comment on the narrative actions by counterpointing radio dramas with the onscreen events.

This effect maps itself onto the romantic trajectory of the two main characters: Bowie (Keith Carradine), a fledgling bank robber, and Keechie (Shelley Duvall). Despite the extensive dialogue exchanges between Bowie and his partners Chickamaw (John Schuck) and T-Dub (Bert Remsen), much of the courtship between Bowie and Keechie is narrated by them listening to radio broadcasts. When Bowie is badly injured in a car accident, he is nursed back to health by Keechie in a room wallpapered with sheet music as the couple listens to a soap-opera adaptation of *Romeo and Juliet*. As Robert Kolker observed, "The radio commentary mocks the couple but makes their adolescent passion the more endearing at the same time. Even more, it refuses to let them alone."[19] Additional radio programs from Franklin Delano Roosevelt and Father Charles Coughlin show how the broadcast medium was used to persuade the American populace, and the film makes explicit the connection between political rhetoric and commercial radio's promulgation of consumerism as an ideological plank of American life.

Like *McCabe and Mrs. Miller*, *Thieves Like Us* marks a return of location sound recording, and Don H. Matthews's sound mixing carefully blends the dialogue with a sound perspective that preserves the spatial characteristics of the locations. Though most scenes were recorded with a boom mike, Matthews also used planted microphones to facilitate cinematographer Jean Boffety's zoom lens cinematography. In the process, however, Altman was still limited by the Hollywood sound recording practices that favored single voices speaking separately rather than overlapping. To achieve his goal of having multiple characters speaking at the same time by recording each character separately, Altman had to look outside the film industry to technology and techniques from the music recording industry.

California Split (1974)

Even though *California Split* does not have a reputation as one of Altman's masterworks, it marks his first concerted effort to move away from sound experimentation in his early 1970s films to a new strategy that marked his productions over the rest of the decade. This involved the use of radio-transmitted lavalier microphones and a multitrack audio recorder to "un-mix" the production dialogue tracks, which allowed Altman to remix them in post-production. On all of his previous films, Altman relied on the attentive ear of the production sound mixer to determine what would be heard on the location audio recordings. Several microphone signals might have been fed to the mixer on set, but the mixer made choices about which voices were recorded on the monophonic dialogue track and in what ratio. Thus from several microphone lines the recording mixer had the final choice about which ones were recorded along with the image. Yet for *California Split* Altman hired Jim Webb, along with his assistants Chris McLaughlan and George Wycoff, who presented Altman with an alternative method for recording multiple synchronous lines of dialogue on location.

Webb was sought out after he developed a technique for recording multiple audio signals in sync on the concert films *Mad Dogs & Englishmen* (Pierre Adidge, 1971) and *Elvis on Tour* (Pierre Adidge and Robert Abel, 1972). Utilizing a modified Stevens Electronics 8-Track Recorder, Webb was able to record seven separate channels of dialogue on location with the eighth channel being reserved for a synchronization tone. He explained that Altman outlined four goals for the equipment: to accommodate improvisation, to let lines of dialogue overlap, to record offscreen dialogue, and to capture the subconversations happening at the same time as the main dialogue.[20] Webb's contribution to Altman's sound techniques came after several earlier efforts to achieve the desired effects from a multitrack recorder:

> When I got to it I was about the third guy he asked to try to put this thing together. . . . But they didn't have any real ideas about how to utilize it. So when I came on the thought that occurred to me was that you can't use open mikes simply because the backgrounds are going to add up and beat you to death. So I said the only way that this is going to be specific enough to use anything is you're going to have to use radios on the dialogue. And so that was my contribution, the technique of using radio mikes.[21]

By combining the multitrack recorder with several tracks fed from wireless lavalier microphones, Webb and Altman were able to reinvent not only how dialogue was recorded on location but also how it was used in films. Similar to the use of multiple cameras to record different angles on a scene simultaneously,

FIGURE 13. The use of the Stevens Electronics 8-Track Recorder on *California Split* (Robert Altman, 1974) let the director record the foreground dialogue between Charlie (Elliott Gould) and Bill (George Segal) as well as the background subconversations.

the multiple microphone inputs and multitrack recorder made it possible for Altman to catalogue and choose the multiple conversations and voices in post-production. This meant that the foreground conversations of the main characters were not always the ones heard in the final mix. Instead, Altman could shift the audience's attention to the background conversations, and reviewer Bruce Berman described its effect as follows: "The sound of cards being shuffled, the whir of the roulette wheel, the click of the chips, the tinkle of the glasses, the lighting, the conversation, the violence and the abominable yet somehow attractive Americanism that surrounds gambling and the consciousness it perpetuates is [*California*] *Split*'s sensual mainstay. . . . The background conversations, made subtly manifest by way of exemplary sound work, contribute strangely integral fragments that tell of junkies in need of fixes and prostitutes on the prowl."[22] Like their use in *McCabe and Mrs. Miller*, the background conversations shed light on the histories and motivations of the central characters Charlie (Elliott Gould) and Bill (George Segal), and created a more realistic depiction of their world (see figure 13).

The lavalier microphones and radio transmitters made it possible for Webb to record several conversations at the same time, but they also introduced a curious effect. Unlike Altman's previous films, which featured an uneasy tension between the sound perspective of dialogue recorded on location with a boom microphone and the insular quality of closely miked looped dialogue, *California Split* merged the two aesthetics. At the same time that Webb's multitrack system for recording multiple conversations and voices constructed a newfound realism in the film, the use of lavalier microphones stripped the voices of their spatial characteristics. While this is not an unusual byproduct, one that was often corrected by the introduction of small amounts of reverberation in post-production mixing, Altman chose to leave the voices "dry" without any added effects. According to Jim Webb: "This format kills perspective. With multitrack, and its inherent use of radio mikes, each separate dialogue track has excellent quality. . . . What is lost is a feeling for the perspective of the environment. And the perspective is the mood whether it be an echoing hallway or a noisy street. It is the thing that gives life to the track. Without it, the movie no longer sounds like it looks."[23] Webb's objections are understandable because they break with an ontological tradition within cinema that dictated that the sound characteristics of the voices should match the spatial characteristics of the location. Yet, as Rick Altman demonstrated, the match between sound scale and image scale started to be broken in the name of dialogue intelligibility very soon after the transition to sound period,[24] and Robert Altman was simply choosing one code of representation over another.

These multiple conversations and sound effects redirect audience attention and give them a broader perspective on the diegetic world of the film and the characters that inhabit it. Also, as in *The Long Goodbye* and *Thieves Like Us*, there is no instrumental score in *California Split*. Instead it features a number of songs from vocalist and pianist Phyllis Shotwell that are heard throughout the film. Unlike the sourced music in *The Long Goodbye* and the diegetic radio broadcasts in *Thieves Like Us*, Shotwell's songs carry over scenes and function "as a commentary on the action."[25] What can be perceived as a form of nondiegetic music early in the film shifts its status as the film progresses. Robert Kolker identified this shift when, "at a certain point in the film songs by Phyllis Shotwell, who will later appear as an entertainer in a Reno casino, suddenly begin to be heard on the sound track, commenting on the action. At one point, when Charlie is crossing a street in Reno, humming to himself, something he sings suddenly merges with a song on the sound track."[26] Instead of remaining outside the narrative Shotwell's songs become meta-diegetic commentaries on the characters and their actions. By the end of the film, when Bill has won a substantial sum gambling yet feels nothing in relation to his success, he leaves the casino and Charlie is heard singing "Bye, Bye Blackbird" with Shotwell on the sound track. At this point the music carries on as meta-diegetic commentary, reflecting on both Bill's and Charlie's departures from the story, even as the film credits begin to roll.[27]

Nashville (1975)

While most cinematic narratives function as translations of preexisting textual narratives, where the script determines the major story contours prior to filming, there were a growing number of cinematic narratives in the 1970s that did not fit with the idea of scripted or scriptable stories. These films developed out of loosely plotted, improvised stories and shifted the main thrust of the narrative from causality to the development and interaction of complex characters. In these films the general uncoupling of the cause-and-effect chain forced the audience to impose new interpretive strategies in order to discern meaningful narrative information. Perhaps the *locus classicus* of this is Robert Altman's *Nashville*.

In *Nashville* Altman used Webb's multitrack location sound recording system to create a new form of nonhierarchical narrative where all his characters had an equal possibility to control the story flow. Also, by replicating the multitrack structure of the recording apparatus in the final film he was able to dissolve the industrial/technological divisions between voice, effects, and music, thereby allowing for a more realistic mode of speech in cinema. Unlike traditional mixing practices that positioned vocal intelligibility as sound's principal function, Altman's "democratic" sound allowed for several overlapping voices to be heard at once. It is no coincidence that overlapping dialogue is the most consistent formal attribute found in the films of Robert Altman. Not only was the director interested in portraying a realistic sense of how people speak in everyday life, but his strategy also revealed the highly constructed nature of most cinematic dialogue. In general, dialogue is the primary motivational force in American cinema, and dialogue intelligibility demands that important narrative information be conveyed to the audience without interference. However, in Altman's films, the acoustic interference of overlapping dialogue was not only part of an aesthetic but pointed to a new narrative strategy of multi-focus narration.

The use of multitrack sound recording devices on location allowed Altman to record several tracks simultaneously and to keep the channels separated for post-production mixing. This gave his production team the flexibility to choose from a variety of possible source voices instead of a single premixed take. More importantly, it allowed him to explore a new approach to presenting a plurality of simultaneous voices. Characters in *Nashville* all have an equal capacity to make their words heard in the production process. The story was built from the multiple character conversations, which meant that the sound track, assembled out of several different tracks and takes, became the foundation for editing the images. The resulting design of the film's sound was the product of a participatory effort on all parts rather than a routine effect that precipitated from the vococentric demands of the narrative.

The post-production side of *Nashville* proved to be somewhat more complicated due to the numerous tracks available to editor Sid Levin and re-recording mixer Richard Portman. This meant that decisions often had to be made between Altman, Levin, Portman, Webb, and screenwriter Joan Tewkesbury regarding what tracks to select and mix. As a result Levin decided that "the soundtrack almost had to be edited first to get the right dialogue and then the picture cut to match."[28] Instead of the script governing the continuity of the shots, the continuity of the audio tracks dictated which images would be used and how they would be edited. No longer hindered by the presence of boom microphones, the actors could be covered in long shot with zoom lenses used to direct the attention of the audience on a character rather than the convention of close-up sound. Where several conversations could be heard at once, Levin was able to shape cinematic narrative by using close-ups to highlight one part of a scene while other simultaneous conversations provided alternate narrative pathways. The final cut of *Nashville* placed primacy on the characters' dialogue and the dramatic content of the scenes over narrative teleology.

The general dominance of the narrative and vococentricity over every stage of the recording and mixing marks a hidden ideological premise that subtended American filmmaking. By critiquing the centrality of the voice in cinema, several observations can be made about the representational practices contained in the cinematic process. In particular, the notion of a central speaking voice and the dominant dialogue of the main characters reveals the conflict between the "ideology of progress" and "a priori optimism" governing the action of most Hollywood characters and the collective goals of the diegetic community.[29] Although most Hollywood films contain an illusion of complete intelligibility through the close miking of central characters and the dominance of the dialogue in the mix, they hide the fact that the governing ideological premise required that the narrative, not the voice, take precedence. Therefore the practice of classical dialogue mixing replicated the function of representative democracy: creating the illusion of community through the perpetuation of a central speaking figure and the suppression of other competing voices.

Altman's multi-focus narratives can be examined in light of this observation as a challenge to the paradoxical nature of the democratic enterprise. His characters are all able to speak simultaneously, to express their individual identities, but at a cost: the loss of a central pathway through the narrative. Altman exploited this function of multi-focus narratives to give the audience enough space to analyze this paradox in his stories. *M*A*S*H* exposed the function of field doctors who heal injured soldiers so that they might return to battle, while *Thieves Like Us* and *California Split* revealed the emptiness behind the American dream of instant wealth and success. But nowhere does Altman's multi-focus approach function better than in *Nashville*.

Nashville begins with a fake commercial for the sound track album for the film, featuring songs from each of the twenty-four main characters. Out of the chaos of the commercial a single voice is heard: the sound of the promotional propaganda spewing from the bullhorns mounted on Replacement Party

FIGURE 14. Haven Hamilton (Henry Gibson) records his bicentennial anthem, "Two Hundred Years," in *Nashville* (Robert Altman, 1975).

presidential candidate Hal Philip Walker's campaign van.[30] Even though the van is seen pulling away from the camera in long shot, the sound of the voice is heard clearly and without additional reverberation. This break between sound and image scale serves a double function: it highlights the recorded nature of the sound, both as a prerecorded speech and as the manipulation of the sound on the sound track, and it reveals the central absence of the candidate by evacuating all spatial acoustic cues, an important trope throughout the film.

The film then cuts from the omnipresent voice of the political campaign to a music studio during the recording of the song "Two Hundred Years" by country music legend Haven Hamilton (Henry Gibson). A series of shots breaks down the space of the studio to reveal Haven in a recording booth while the backup singers and musicians are also isolated in their own spaces (see figure 14). Yet, in contrast to the visual separation of the characters, the music heard on the sound track is the completed song mix. Altman begins the film with this unification of singing voices, extolling the national spirit surrounding the bicentennial, as a way of showing the fabricated nature of the recording process. With the unwarranted entrance of Opal (Geraldine Chaplin), a reporter for the BBC, the song structure breaks down and the audience is granted access to each of the individual vocal sources. The hierarchy of the mixing process has given way to a direct democratic practice, one in which any voice can be heard on the sound track, but only after the fictional unity of the song mix is discontinued. Throughout the film the relationship between the structure and technology of music is paralleled with the democratic endeavor through the subplot of the presidential primaries. Moreover, this shift between competing acoustic strategies foregrounds the manipulation inherent in classical dialogue mixing.

Altman's use of Nashville, Tennessee, as a backdrop for a presidential campaign made the democratic function manifest by equating it with the capital of the country music industry. The film oscillates between passages of polyphony, where three and four conversations occur simultaneously, to moments of sonic organization, always portrayed in the form of a song. This oscillation between the two principles of sound mixing mirrors the American crisis of identity in the post-Watergate period, and the crisis is sublimated in *Nashville* through its plurality of voices. Ultimately the characters' identity crises are transferred to the spectator/auditor, who has to choose which elements to listen to and which to ignore. It is only during the musical passages that a false sense of unity is established through the post-production sound mix and the democratic array of competing voices is silenced in favor of one central melodic voice. Thus, in this oscillation between the multiple voices and the songs, the spectator/auditor is made aware of the construction of the sound track and the narrative function is called into question.

Rick Altman made some astute observations about this film and the radical approach to sound mixing in his article "24-Track Narrative? Robert Altman's *Nashville.*" He claims that while the film does contain a radical gesture in its approach to recording and mixing the voices in the film, "this non-hierarchic openness oriented to spectator choice yields in the end to the narrative logic of the traditional linear model."[31] While the film does build to a narrative climax, the assassination of singer Barbara Jean (Ronee Blakley), which functions as the unifying point for the sound track, I believe that a much richer reading and explanation can be gained by looking at the scene as an attempt to restore faith in a vanishing democratic fantasy.

The final sequence of *Nashville* is structured around a political rally for Walker at the Parthenon—Nashville's full-scale replica of the Temple of Athena in Athens, Greece, the birthplace of direct democracy—where country singer Barbara Jean is mysteriously shot by another character, Kenny Fraiser (David Hayward). The ensuing actions are all carefully mixed on the sound track to allow the audience to comprehend the action and to follow the events that unfold. Rick Altman noted that this "unexpected poverty of the sound mix, at the very point when all twenty-four tracks are available for the final scene, can clearly be attributed to narrative imperatives."[32] Although I agree that the film does choose to utilize the sound according to narrative strategies at this point, I argue that it marks a final attempt at creating a false sense of narrative unity, restored in a moment of trauma.

During the final scene the pace of the editing dramatically increases, thereby also mimicking the tension of the characters. Through standard cinematic conventions, such as close-ups of each of the characters and cuts to Barbara Jean singing "My Idaho Home," a dramatic connection between the characters in the audience and Barbara Jean is created. Like a game of Russian roulette, the images cut back and forth until the camera keeps returning to Kenny as he fumbles with the gun in his violin case. This editing pattern is broken by a full-frame close-up of the American flag hanging above the stage as a single wave ripples across its surface. Almost as if this image precipitates the action, Barbara Jean's song ends and Kenny opens fire. The conflation of the symbol of the nation and Barbara Jean reinforces the collapse of narrative unity into the democratic plurality of voices in the film. In her death a sense of "order" is restored as the character Albuquerque (Barbara Harris) takes the microphone and begins to sing the song "It Don't Worry Me." At this moment the entire crowd begins to take up the song and to chant the chorus, "Some may say that I'm not free, but it don't worry me," over and over again as the images fade to black. The song continues over the end titles, refusing to stop for a full five minutes after the credits end.

It is this final moment of the film that fully demonstrates Robert Altman's critique of both the American democratic process and classical narrative form. True democracy is won in his film at the cost of the national endeavor. In order to restore a sense of order and nation, the traumatic event is covered over by the unifying power of song. The film shows the music industry as an analog for the national project where a plurality of voices can be organized into a unified whole only at the loss of the individual character of the voices. Confronted with the trauma of Barbara Jean's shooting, the only way that the crowd can organize themselves and refute the event is by singing this song of denial, what Robert Kolker called "their great anthem of passivity."[33] By letting the song extend beyond the boundaries of the film and permeate the space of the theater, Altman did not yield to the demands of closure but exposed the central function of narrative as a way of assuaging and controlling the spectator/auditor. The extension of the final song over the end credits and into the theater itself cracked open the classical narrative structure to reveal the absence at its core.

Robert Altman's films stand as wrinkles in the smooth fabric of the evolution of sound films. *Nashville* marks the end of a period of experimentation in cinema sound that challenged the hegemony of the hierarchic structure of sound production, a brief moment of radical sound use that exploited the capability of recording technology to provide a new model for sound reproduction. What is at stake here is the fact that cinematic narrative theory draws its conclusions from models developed in classical Hollywood cinema and reified during the period after the introduction of Dolby Stereo. In general, narrative theories do not examine the richly experimental, albeit brief, period of the 1970s, and none investigate the industrial/technological biases imposed on narrative by the cinematic process. By exploring the films of Robert Altman from the 1970s it is possible to discover new insights to allow for a rethinking of the function of narrative in cinema.

7

Martin Scorsese's Dialectical Sound

I discovered rock 'n' roll in 1956—real rock 'n' roll, Little Richard, Fats Domino, Chuck Berry, Elvis, Screamin' Jay Hawkins. For me, it was a real revolution.

—Martin Scorsese

Along with directors who came from television or exploitation cinema, another group came from film schools and brought differing sensibilities to film sound strategies. In particular, directors like Martin Scorsese emphasized the importance of popular music while also introducing new elements of acoustic realism into their films. Scorsese qualifies as a sound auteur in terms of both his meticulous sound choices and the complicated musical structures of his sound tracks. His active participation during all phases of post-production sound editing and re-recording ensured a consistency of sound use from film to film despite many changes in personnel.[1] In addition, his work as an editor on *Woodstock* (Michael Wadleigh, 1970) and *Medicine Ball Caravan* (François Reichenbach, 1971) and as sound effects editor for *Minnie and Moskovitz* (John Cassavetes, 1971) tuned his sensibility to the ways in which sound and music could aid the storytelling process.[2] By his second feature film, *Boxcar Bertha* (1972), some of his early experiments in sound juxtaposition and layering can be heard. The opening sequence was constructed entirely in the studio where looped dialogue, railroad sledgehammers, a harmonica, and a crop duster all interweave in a complex audio montage. Scorsese also experimented with associative editing based on sounds when the death of Bertha's father, registered by her scream, dissolves into a train whistle in a Hitchcockian allusion. There are several points in the film where the placement of the microphone in a long shot—such as the sound of fiddle players initially heard as nondiegetic until the camera tracks to follow character movement only to reveal the sound source, or the use of offscreen sound motifs like the train whistle or the clank of the freight cars—draws the attention of the audience to certain characters or activities not revealed in the image. In part this may be a felicity of low-budget filmmaking, where it is easier to allude to the presence of offscreen elements than to show them, but it is a strategy that carries over to his later films as well.

From the beginning Scorsese was aware of the potential of prerecorded score music, and he has successfully explored several models of its use. Unlike many of his contemporaries, Scorsese was not interested in using music to produce a superficial nostalgia in his films; rather, he was more concerned with the dialectical relationship between the music, character portrayal, and the development of the story. While Lucas was able to create a partially dialectical relationship between the music and story in *American Graffiti* (1973), primarily through Walter Murch's careful manipulation of the songs to foreground their diegetic presence,[3] Scorsese's films construct the music as a complex intertext for the onscreen events. Through a conjunction between musical styles, genres, and models of their diegetic use, Scorsese created a double articulation of meaning in his films.

In this regard Scorsese's films have only one true predecessor in their dialectical use of the compilation score: Kenneth Anger's 1964 independent short *Scorpio Rising*. Anger's film is unique in its strict separation between the sound track and the images due to the film being shot silent, with all music and sound effects added in post-production. Although it is clear to the audience that the sound and images are neither synchronous nor ontologically linked, due to the absence of any possible sources for the music, there is a deliberate correspondence between the music and the narrative events of the film. Anger built the sound track out of thirteen contemporary pop songs,[4] stretching as far back as Ray Charles's 1961 version of "Hit the Road Jack." Each song is heard in its entirety and the songs feature loose lyrical connections with the onscreen events. Through the strict structural separation between the sound track and the images, the spectator is forced into a critical perspective on the film and asked to question the working relationship between the music and the narrative. As Ed Lowry pointed out, the film is viewed as an open text, one in "which every signifier is cut loose from its culturally normative signified and presented as simultaneously comical and dangerous."[5] By doing so Anger was able to create a polemical relationship between both the songs and the images as well as between the songs themselves. Anger not only captured the relationship between popular music and the homosexuality and ritualistic fetishism of the film, but he also set up a series of questions about the internal logic of the song choices themselves. The horizontal relationship between Gene McDaniels's "Point of No Return" and Little Peggy March's "I Will Follow Him" foregrounds the nihilism in the lyrical content of the songs. This relationship is related to the vertical relationship between "Point of No Return" and a motorcycle race, and how "I Will Follow Him" accompanies images of Scorpio intercut with those of Hitler and Christ. In this way Anger set up a two-tiered system of interactions: a horizontal relationship between the songs as a group and a vertical relationship between the images and the songs. Scorsese borrowed this strategy for his films in the 1970s and refined it to a very precise system of signification.[6]

Scorsese worked out a number of his musical strategies in his first feature film, *Who's That Knocking at My Door?* Shot from 1965 to 1966 and initially released as *I Call First* in 1967, *Who's That Knocking at My Door?* did not receive national distribution until 1968. The film, set in the first half of the 1960s, features several doo-wop songs that accompany the central Italian American character, J.R. (Harvey Keitel), and work to establish the period of the film. However, the familiar accompaniment of the doo-wop songs is placed in counterpoint with two rock numbers. The first rock song, Mitch Ryder and the Detroit Wheels' "Jenny Take a Ride," accompanies the opening scene where J.R. and his friends are seen viciously beating a rival street gang. The second, The Doors' "The End," accompanies J.R. on a fantasy sequence where he passionately makes love to a number of anonymous women. This correlation between the use of rock music and acts of physical or emotional violence carries over directly to *Mean Streets*, where Scorsese developed a syntax of musical choices to reflect a higher order signification. The salient difference between the two films is that the use of doo-wop music in *Mean Streets*, a film set in 1973, stands out as distinctly anachronistic.

Mean Streets (1973)

In *Mean Streets* Scorsese used his song choices to create a secondary meaning-bearing structure through which he could engage audience knowledge external to the narrative. Generally, the song choices are broken down into distinct genres: doo-wop songs from the early 1960s, rock songs from the early 1970s, and Neapolitan folk songs. Though these genres may seem to be mutually exclusive, it should be pointed out that the doo-wop songs from the early 1960s were a nascent combination of rock 'n' roll instrumentation and rhythm and blues (R&B) harmonies. Just as rock 'n' roll embodied a number of different styles before it was distilled to the almost exclusively white, guitar-based rock of the late 1960s, doo-wop was more of an urban phenomenon rather than an exclusively black form of music. Several Italian American groups such as Dion and the Belmonts, The Capris, Danny & the Juniors, and The Passions were heard on early 1960s radio along with their African American counterparts The Shells, The Cadillacs, The Orioles, The Del-Vikings, and The Jesters. Racial differences were not a major factor as evidenced by a number of mixed-race groups, including

Johnny Maestro and the Crests, The Impalas, and the Italian-themed African American group Little Caesar and the Romans. However, in the late 1960s, with rock 'n' roll becoming rock and R&B changing into the newly christened soul music, these divergent musical divisions were split along racial lines with rock being predominantly white and soul almost exclusively black. Scorsese's musical choices in *Mean Streets* knowingly activated questions of racial difference after this paradigmatic musical shift, where every musical choice contains one level of signification in regard to its content and use and a second level in regard to its style and associated cultural implications.

Robert Kolker identified the narrative function of this "two-tiered system" in *Mean Streets*, pointing out that "Scorsese carefully integrates a double perspective in the film, a free-flowing observation and a carefully structured point of view both of and from a central character."[7] This internal and external perspective is doubly articulated through two central devices in the film: the internal monologues of the main character, Charlie (Harvey Keitel), and the interplay between the song choices and the diegetic action. The contours of the story are relatively simple: Charlie is caught between being groomed to take over his uncle Giovanni's restaurant while also looking out for his loose-cannon friend, Johnny Boy (Robert De Niro). Scorsese guides the audience through Charlie's actions, but the director also uses voiceover to grant the audience access to Charlie's thoughts. Charlie's voice is the first thing heard in the film as he intones over a black screen: "You don't make up for your sins in church; you do it in the streets; you do it at home—the rest is bullshit and you know it." The montage of Super-8 footage that follows provides the audience with visual information in the form of memories that correspond to the character's internal narration. In a later scene, however, Charlie enters a church and kneels before the altar as a voice is heard saying, "Lord, I'm not worthy to eat your flesh—not worthy to drink your blood." The audience initially assumes this to be Charlie's internal voice, but as he speaks the final phrase aloud, we realize that the first voice was another's, the voice of Scorsese himself. In this way Scorsese simultaneously positions Charlie as a surrogate for the director in the film while also introducing a central question regarding the authenticity of sound use in the film. Specifically, the audience is unsure whether they are supposed to hear the two voices as different parts of Charlie, internal and external, or whether they are meant to hear the first voice as the voice of the director, standing offscreen, feeding the lines to his actor.

By constructing uneasiness within the audience from the first scenes, the film prompts them to listen carefully to the acoustic choices and to pay attention to how their structure inflects the film's overall meaning. After the film starts with Charlie's voice over a black screen, it switches to an image of him waking up and moving around his bedroom accompanied by the ambient sounds of his apartment building and the street below. When he returns to bed, four quick cuts emphasize his head hitting the pillow as we hear the introductory drum flourish of The Ronettes' "Be My Baby." From this first instance the audience is asked to regard not only the lyrical content and genre of the music but also the rhythmic and stylistic ways it relates to the visuals. When Charlie arrives at his friend Tony's club in the following scene, "Tell Me" by The Rolling Stones is playing over the speakers. Although it is clear that the music has a diegetic source and that several people are dancing along with it, there are elements that do not fit with the images. First, Charlie's entrance to the club is shot from his point of view in slow motion. Over the images we hear his voiceover narration and the opening guitar chords from the song with no added reverberation, making it seem as though the music is internal to Charlie. Only after the scene cuts to an establishing shot and returns to normal speed does the spectator realize that the music is actually being played in the club. This is because the characters react to the beat of the music, not because there is any shift in acoustic spatialization as the song plays through to its conclusion.

Derek and the Dominos' "I Looked Away" starts immediately thereafter, establishing a jukebox logic that is in complete accord with the sounds of the club. However, with the entrance of Johnny Boy, the music fades away and Charlie's internal monologue returns, saying, "All right, OK, thanks a lot, Lord, thanks a lot for opening my eyes—we talk about penance and you send this through the door." Immediately the visual perspective of the film shifts to Johnny Boy's point of view, replicating Charlie's prior entrance, as "Jumpin' Jack Flash" starts on the sound track (see figure 15). Again, the use of slow motion over another Rolling Stones song creates a strong parallel between Johnny Boy

FIGURE 15. Johnny Boy's (Robert De Niro) flamboyant entrance in *Mean Streets* (Martin Scorsese, 1973) is accompanied by the Rolling Stones' "Jumpin' Jack Flash."

and Charlie and their privileged relationship to the music. Yet with this second entrance the slow motion continues for forty seconds with the regular noises of the bar slowed down and reduced in volume. The dreamlike effect of Johnny Boy's entrance is enhanced by the fact that the music seems to be external to the space but internal to his character. As Ian Penman noted, "It's almost as if De Niro carries the music around with him (in his hips, head, angles, jerky movements, gum chewing)."[8] Hence the music is constructed as an extension of the internal psychology of the characters instead of as an external narrative agent. Importantly, the contrast between the R&B flavored "Tell Me" (from 1964) and the straight rock of "Jumpin' Jack Flash" (from 1968) sets up a crucial difference between Charlie and Johnny Boy that is played out in both the story and on the sound track.

Moving beyond the lyrical relationship between the song and plot development, Scorsese placed the songs into a dialectical relationship with the characters and their actions by constantly shifting the ontological status of the music. Often the songs are sourced, whether through car radios, juke boxes, or the public address system at Tony's bar, which provides a solid connection between the music and its presumed point of origin. However, in nearly every scene where a source for the music is clearly identified, Scorsese manipulated the sound to cast the music's diegetic origin into doubt. In fact, several songs shift between the diegetic and nondiegetic registers to create a complex system of interaction between individual characters and the song structures. Using music as sound bridges over narrative ellipses and jump cuts, shifts in timbre and volume to indicate subjective changes, and differences between real time and musical time, the film forces the audience to question the nature and significance of the musical choices. Whereas most compilation scores parallel the narrative events and reinforce the story details,[9] *Mean Streets* uses music to provide diegetic verisimilitude while simultaneously engaging the audience in a process of decoding secondary meanings. Through this technique Scorsese introduced an additional layer of commentary on the characters and their actions. In many ways the slippage between diegetically sourced songs and meta-diegetic internal music[10] makes the spectator aware of how the music has less to do with story than with characterization. Pauline Kael described this effect best: "*Mean Streets* doesn't use music, as *Easy Rider* sometimes did, to do the movie's work for it. . . . The music here isn't our music, meant to put us in the mood of the movie, but the characters' music."[11] Often it becomes an internal element of the characters, following them across cuts and scene changes as if the music were providing accompaniment to a character's actions.

The music provides a link between the internal subjectivity of the characters and the environment in which they live while also creating a structure that emulates the narrative events within the film. Charlie is caught between the traditions of his Italian American community and a desire to experience his life through events and actions that are rejected by his family and friends. Specifically, questions of race and identity enter into the film through Charlie's interaction with several characters outside his hermetic community. During the introductory scene in Tony's club, Charlie is infatuated with Diane (Jeannie Bell), an African American stripper, but he is unable to act on his attraction. The fact that the scene plays out with Diane dancing to two rock songs, "Tell Me" and "I Looked Away," emphasizes how the mixed racial components of doo-wop from the early 1960s changed to nearly exclusively white rock by the 1970s. The opposition between Diane and the music is foregrounded when Charlie notes, in voiceover, "She is really good looking . . . but she's black." This ethnic and racial distinction also forces the spectator to listen for similar difference in the musical choices. Scorsese actively uses these choices to emphasize the racial contradictions within the film. Most powerfully, as the music oscillates between the mixed racial component of early 1960s doo-wop and rock, a framework of violence is exposed.

Because the Italian American characters in the film are forced into a refusal of other national cultures, religions, and races, the unstated issues of intolerance manifest themselves first on the sound track and second in the form of sporadic eruptions of violence. A pattern emerges across the film that contrasts the racial stability of doo-wop music with exclusively white rock and Italian contemporary songs. In general, doo-wop music is used to augment many of the comic scenes while the eruptions of violence are nearly always accompanied by rock or Italian music. Doo-wop songs "Desiree" by The Charts and "Oldies but Goodies" by Little Caesar and the Romans play over the scene where Johnny Boy first underpays his loan to Michael (Richard Romanus), while The Shirelles' "I Met Him on a Sunday" is heard when Michael and Tony (David Proval) "take the kids for $20."[12] Although there are no attempts to connect the songs' lyrical contents to diegetic actions, as in *Scorpio Rising*, there is nonetheless an intrinsic connection between the styles of music and the narrative events. Even when the film works against these internal paradigms, as when the fight breaks out in the poolroom to The Marvelettes' "Please Mr. Postman," it is in order to assure the spectators that the violence in the scene is played for comedy. This paradigm is contrasted by the use of Jimmy Rosselli's "Mala Femmena" and Renato Carosone's "Maruzzella" during the hit at Tony's bar.[13] The Italian songs are out of place in Tony's club, and by breaking the pattern of rock or doo-wop it is as though the displacement of music precipitates the killing.

This disruption of the established pattern and eruption of violence creates a ripple effect throughout. The film appears to return to a state of equilibrium when a party is thrown at Tony's bar for Jerry (Harry Northrup), a returning Vietnam War veteran. As Tony urges Michael to "play only oldies" on the jukebox, several curious things happen. First, as the characters get more intoxicated, the visual strategies of the film change to emphasize their disorientation. This culminates in a shot where the camera apparatus is physically mounted to Harvey Keitel's body while the Chips' "Rubber Biscuit" is heard playing. The use of the camera mount creates a powerful sense of subjectivity where the audience sees everything moving around Charlie as his face and torso remain absolutely stationary within the frame, even after he passes out on the floor. To reinforce this subjectivity and to emphasize Charlie's intoxication, the Chips' song is heard directly, with no added reverberation or spatial characteristics, while all the other noises of the bar are eliminated. Once again, the break in the film's strict pattern of adding spatial characteristics and reverberation to sourced music is coincident with Jerry's violent outburst when he attacks a girl at the bar without provocation. This action is accompanied by Johnny Ace's 1955 R&B hit "Pledging My Love,"[14] and from this point the spectator is kept in suspense because the secure divisions between musical genres and narrative action are cast into doubt.

Throughout most of the remaining scenes Scorsese exclusively used the diegetic Italian folk music of the San Gennaro street festival. Not only did this set up a system of musical stasis, being neither rock nor doo-wop, but it also firmly repositioned the characters in the temporal flow of the festival and makes the audience aware that the prior events transpired within the time frame of a little over a week. Moreover, the use of the Italian folk music provided a background against which the presence of other genres became more legible. When Charlie convinces Johnny Boy that he has to leave town to avoid Michael, the presence of "Mickey's Monkey" by The Miracles on the car radio leads the audience to believe that the two can escape safely. Yet the extreme violence of the film's penultimate scene, where Michael and his hired gun (played by Scorsese) track down and shoot Johnny Boy, is presaged by the presence of "Steppin' Out" by Cream from *Live Cream, Volume Two*.[15] With Eric Clapton's fevered guitar solo as accompaniment, this last act of violence firmly attaches itself to the separation of rock music from R&B and the subsequent cultural split between the two genres. The final correlation between acts of extreme violence and rock music retrospectively creates an irreparable split in the film as the racial and cultural separation of the characters is mapped onto the musical shift from doo-wop to the break between rock and soul in the late 1960s.

A review of the production history surrounding *Mean Streets* makes it even more impressive that Martin Scorsese's first film for a major studio is such an intensely personal and original film. The story is set in New York's Little Italy during the two-week Feast of San Gennaro, a public street festival held every year during the middle of September. Aside from a few exterior shots, however, nearly all of the film was made in Hollywood with the streets of downtown Los Angeles doubling for New York.[16] Because Warner Bros. forced Scorsese to shoot in Los Angeles, there are a number of inconsistencies in lighting, set design, and particularly sound quality. Many of the conversations in the film take place at a very rapid pace with several characters talking at once, yet unlike Robert Altman's sound tracks, where the recording and mixing of the voices allows for most of the lines to be heard, often Scorsese's dialogue becomes an aggregate of mumbled voices and slang. This is not meant as a negative assessment, however, since the multiple voices on the production recording made it difficult to edit and shift lines in post-production. As a result, there are a number of scenes where poor production dialogue and crackling radio microphones are used, but there are as many scenes where all the dialogue was looped and replaced in post-production. Unlike many of his New York counterparts, Scorsese admitted that he liked looping to clarify certain story points, but in doing so something was lost in the inflection and feeling of the lines.[17] The result is that the film has a somewhat uneven and rough sense of dialogue that is often inconsistent from scene to scene. But Scorsese took what would be commonly heard as a distraction in his film and turned it into a central structural aspect for making the audience aware of the different functions of the voice in cinema.

With its release on 2 October 1973, *Mean Streets* signaled the presence of Scorsese's personal voice through his dialectically constructed sound track. The music in the film did more than just hide the deficiencies in the sound track from a limited production budget and location shooting; it was structured to create a second level of meaning in the film. More than any other director in the 1970s, Scorsese was able to use music to create a double articulation of meaning that allows spectator access to character psychologies while also working as a schema for understanding the ramifications of their actions. By tapping into two separate musical functions—both unifying and disruptive—Scorsese elevated the sound track from a level of simple commentary to a complex structural interaction with both the narrative events and the larger cultural context of the characters and their world.

Alice Doesn't Live Here Anymore (1974)

Mean Streets brought Scorsese a large measure of success as well as a slew of Mafia-related scripts in the wake of *The Godfather*'s success. Yet instead of embarking on another investigation of Italian American life, Ellen Burstyn sought out Scorsese to direct *Alice Doesn't Live Here Anymore* from Robert Getchell's script about a widow trying to reinvent her life while taking care of her teenage son. The film moves from Socorro, New Mexico, to Phoenix and eventually Tucson, Arizona—a world away from Scorsese's New York. Despite being an unlikely follow-up, *Alice* followed several of the same sonic and musical strategies used in *Mean Streets*. The music in the film—the songs performed by Alice (Ellen Burstyn) and the diegetic songs heard through a variety of sources—all adhered to the same two-tiered system of signification. Moreover, Scorsese carefully balanced the use of dialogue and sound effects to develop a verisimilitude of the locations and the character emotions as they traversed the southwest. Unlike Coppola and *The Rain People* (1969), Scorsese was not interested in expressing a subjective sense of character. Instead he preferred to use sound to reinforce the realism of the locations and temporality with an emphasis on character voices and dialogue.

Dialogue is central to the story in *Alice* and it dominates the film's mix. That is not to say that the film is less original than *Mean Streets* but that it deploys its audiovisual strategies in a less overt fashion. Whereas *Mean Streets* had fixed and stable camera placements, *Alice* keeps the camera in motion, giving it a dynamism that contrasts with Alice's trapped life as a housewife. With the many complex tracking shots involved in the filming, often location sound recording was compromised and Scorsese looped much of the dialogue. This creates a curiously intimate effect that draws the spectator closer to the characters, which contrasts with the veil of reverberation heard in location dialogue from *The Rain People* that keeps spectators at more of a distance from the diegesis. Production records show that dialogue replacement began with the principals shortly after the end of production in late August of 1974 and concluded in mid-October with many of the secondary characters.[18] The lengthy audio post-production schedule hints at the careful attention paid to dialogue in the film and its significance in the storytelling process. Indeed, Scorsese worked with Burstyn and the other actors in pre-production to improvise interactions and new dialogue. Similar to John Schlesinger's method on *Midnight Cowboy* (1969), these improvs were recorded on audiotape and given to screenwriter Robert Getchell to include in the final script.[19]

Another major difference is that, unlike with Charlie in *Mean Streets*, we are not given direct access to Alice's thoughts through voiceover. Instead, the script includes several comments from Alice that are delivered almost as theatrical asides, where it is unclear whether she is talking to herself or directly to the audience. Though these moments never threaten to break the cinematic fourth wall, the technique grants the audience access to her frustrations while also supporting much of the film's comedy. Much like Elliott Gould's portrayal of Philip Marlowe in Robert Altman's *The Long Goodbye* (1973), Alice's running commentary is more for the external audience than for the diegetic characters. Alice's asides also make her much more endearing to an audience that can sympathize with her situation via the lens of comedy. In accordance with the shift between direct dialogue and Alice's sotto voce comments, the dialogue is carefully mixed to fade in and out within shots, emphasizing Alice's lines while occasionally deemphasizing those of other characters. This is part of an expressive use of sound in the film that serves to keep the viewer close to Alice and telegraphs her emotions through the shifting soundscape.

We are cued to contrast Alice's present with her fairy-tale memories of childhood. In a lush introductory sequence that harkens back to the Technicolor melodramas and musicals of the 1940s, a young Alice is seen "singing 'You'll Never Know How Much I Love You' from *Hello, Frisco, Hello* before a set which is a direct and expensive homage to William Cameron Menzies."[20] The music, which begins as nondiegetic over the title credits, continues as the Academy ratio image dissolves to a soundstage set that presents Alice's childhood in Monterey, California, as a stylized movie memory: part *The Wizard of Oz*, part *Gone with the Wind.* Strolling along a dirt path the young Alice pauses as the nondiegetic song ends and she proclaims, "Wait. No, wait. I can do that better." Her voice, however, is that of an adolescent girl and her off-key warbling contrasts sharply with her imagined talent. The dialogue breaks with the scene's artifice when Alice muses, "I can sing better than Alice Faye, I swear to Christ I can," only to have her mother interrupt the idyll, shouting, "You get in this house before I beat the living daylights out of you." Alice's swearing and her mother's brash statement are shockingly inappropriate for the genre and time period being evoked, but Alice's subsequent dialogue reveals her deeply rooted desires and drive. "You wait and see," she confides to her doll, "and if anybody doesn't like it they can blow it out their ass."

This opening sequence is extremely important in constructing Alice's character because it sets up several of the main tropes that follow. First, it establishes her longstanding desire to be a singer and a tenacity that outweighs her abilities. Second, because the rest of the film is shot on location and foregrounds its contemporary realism, the introductory sequence cannot be seen as historical reality but needs to be read as Alice's memory colored by the Hollywood films that formed her worldview as a child. And, third, it introduces the device of Alice's spoken asides that reveal her socially inappropriate comments bordering on coprolalia. The opening sequence ends with a shocking audio effect when the last word of Alice's rendition of "You'll Never Know How Much I Love You"—"NOW!"—shudders, repeats, and increases in volume as she runs into her house. Triggered by the sound of the slamming front door, the image shrinks on the screen and the sound of a jet taking off creates an abrupt audio transition to the present. As Karyn Kay and Gerald Peary described it: "A tornado-like camera movement propels Alice out of her childhood. The camera swirls high above the farm and comes down to rest in an alien time and place: modern suburbia, where the hard rock unisex of Mott the Hoople blares from the soundtrack with gale force. Glitter fey replaces Alice Faye."[21] As the

FIGURE 16. Alice's reminiscences of movie musicals in *Alice Doesn't Live Here Anymore* (Martin Scorsese, 1974) are literally drowned out by Tommy (Alfred Lutter) grooving to Mott the Hoople's "All the Way from Memphis."

band kicks into their 1973 hit "All the Way from Memphis," a long-take crane shot tracks down a suburban Socorro street to reveal Alice's house with her in the sewing room mouthing the words to a different song. Where the opening sequence included a nondiegetic song that was linked to Alice's childhood memories, the Mott the Hoople song that begins as nondiegetic is revealed to be sourced, coming from Alice's son Tommy (Alfred Lutter) listening to records in the living room (see figure 16). The parallelism of the music reveals that in the twenty-seven-year gap between the opening and the present Alice's dreams of becoming a singer have gone unfulfilled.

The stark contrast between the 1970s rock songs heard throughout the film and Alice's gauzy renditions of 1940s torch songs reveals the latter to be a nostalgia for an imagined past and an idealized present. In fact, even the film's title comes from the "old Carmen Lombardo standard" "Annie Doesn't Live Here Any More," written by Joe Young, Johnny Burke, and Harold Spina in 1933.[22] Scorsese commented on the inclusion of both contemporary rock and 1940s popular songs: "The choices of music came from the characters' heads. In other words, the kid listened to rock 'n' roll because I felt this kid would listen to Mott the Hoople, Elton John, and Leon Russell. . . . And we really had to think about the songs that they as characters would listen to."[23] This metaphoric use of songs as the extensions of the characters carries over its use from *Mean Streets* and points forward to the literal idea of music in a character's head in *Taxi Driver*.

The sharp division between the songs of Alice's past and those of Tommy's present indicates the significance of music in shaping and defining generations. As Scorsese commented in an interview with Steve Howard, "We had to be very careful about songs commenting on the action and really stick with things which would bring out different aspects of the character."[24] This linkage of character and music is a trope that carries on through most of Scorsese's future films and which has its foundation in *Mean Streets* and *Who's That Knocking at My Door?*

Another important sound practice that returned in *Alice Doesn't Live Here Anymore* was the use of offscreen sound effects to extend the diegesis while also offering metaphorical commentary on the story. There are many times in the film when offscreen sounds are heard intruding on Alice and Tommy's world. While some of this may be the result of shooting on location in Arizona and New Mexico—Scorsese discussed the problems of ambient street noise and the difficulty of using radio microphones instead of booms[25]—the addition of offscreen sounds developed into strategy for commenting on the story. In Socorro, the jet plane sound that transitions from the credit sequence recurs multiple times to remind us of the presence of the nearby Holloman Air Force Base as well as to introduce an unsettling audio element before the death of Alice's husband, Donald (Billy Green Bush). As Alice and Tommy travel across the country they are often subjected to sounds intruding from outside their hotel rooms. The first night Alice sleeps with Ben (Harvey Keitel), a younger man who picked her up at her job singing in a nightclub, the sounds of breaking bottles and distant sirens are heard outside their room. After a dissolve to Alice back in bed with Tommy in their hotel, they are awakened by a violent domestic dispute in the next room. The sounds of violence serve as harbingers of Ben's hidden brutality that eventually surface and force Alice and Tommy to leave Phoenix abruptly.

When in Tucson, Alice finds work as a waitress at Mel & Ruby's Café, yet the sounds of her travel deferred continue to follow her. Russell Davis described her life in Tucson as follows: "The sun boils down from an impersonal sky. The frame widens to take in garbage cans, then the desert, and in the distance the crouching, ever-present mountains. The soundtrack expands with the buzzing of continuous traffic on a hot pavement."[26] The external sounds of the road intrude on Alice's world as a reminder of her goal of returning to Monterrey with Tommy. Scorsese's use of these unseen sound sources reveals his technique of using sound effects to reinforce the diegetic realism while also commenting on the characters and story. The experiments in the dialectical use of sound and music in *Mean Streets* and *Alice Doesn't Live Here Anymore* come to their full expressiveness with the complexly evocative soundscape for *Taxi Driver*.

Taxi Driver (1976)

If the use of diegetic music and sound effects in *Alice Doesn't Live Here Anymore* was designed to help the central character hold together her identity and her relationships, then the dominant aesthetic of *Taxi Driver* is the sound of coming apart. The story of Travis Bickle (Robert De Niro), a disheartened Vietnam vet living in New York and trying to make sense of his world, was written by Paul Schrader and loosely based on the diaries of Arthur Bremer, the man who attempted to assassinate presidential candidate George Wallace in 1972. Travis is a cipher and little background information about him is divulged. When he applies for a job as a cab driver at the beginning of the film, we realize that he's a former marine who was discharged in 1973. Like Ethan Edwards in *The Searchers* (John Ford, 1956)—a film that is a touchstone not only for Scorsese and screenwriter Paul Schrader but for many in the new American cinema of the 1970s[27]—the three years between when he was discharged and the present are a lacuna. We don't know any more of Travis's backstory or why he decided to work as a cab driver. Instead, the information we discover about his character comes from three sources: the sounds and images of his daily routines, his inner thoughts conveyed to his diary and expressed as voiceover, and Bernard Herrmann's musical score.

It is important to point out that *Taxi Driver* was the first film in which Scorsese used a composer and nondiegetic score. Like the use of diegetic source music to set up a two-tiered system of allusion to distinguish one style of music from another in *Mean Streets* and *Alice Doesn't Live Here Anymore*, *Taxi Driver* uses Herrmann's nondiegetic score music to similar effect. Also, as in his previous films, there is an oscillation between interior subjectivity and exterior objectivity; but unlike in the prior films, in *Taxi Driver* that distinction is not always clear. In *Mean Streets* we were let into Charlie's consciousness to hear his internal thoughts as voiceover, yet the narrative was not restricted entirely to his perspective and several scenes occurred where Charlie was not present. *Alice Doesn't Live Here Anymore* had a more restricted narration where Alice was present during most of the scenes, and aside from the opening sequence its presentation was objective. In *Taxi Driver*, the narration is tightly restricted and nearly all the scenes are presented from Travis's perspective. As Michael Dempsey observed, "It's definitely a subjective vision—the film locks us into his consciousness."[28] Yet instead of relying on point of view to convey Travis's perspective on the world, Scorsese lets sound do the bulk of the work. Nearly all the sounds in the film are filtered through Travis's consciousness, and as the soundscape shifts from realistic to restricted to suspended,[29] these changes are read as changes in Travis's mindset.

In general the sound effects and ambiences in the film are used to reinforce and even heighten its gritty realism. Heard from Travis's point of audition in the cab and his apartment, the offscreen sounds of the city regularly bleed through windows and walls and are presented to match their sound sources to Travis's location. There are some sounds, however, that work against this sense of realism and prompt the audience to consider a reading position different from classical realism. Curiously, the sound of the cab Travis drives is generally absent, and the lack of cab sounds makes his nocturnal rounds seem more ethereal. His gliding around the city contrasts greatly with the strict automotive realism of films like *Two-Lane Blacktop*, and the antirealism of Travis's cab is one of many sound strategies that unhinge the standard codes of narrative transparency from classical Hollywood cinema. Scorsese subverted several of these conventions to render Travis's psychopathic mindset through sound.

Following what is arguably the primary trope of film noir, Scorsese's noir redux relies on Travis's voiceover to convey a substantial part of the narrative information. "Travis' diaries are heard in constant voice-overs and act as a kind of mirror to Travis' eye and thoughts, as well as his movements," explained Ann Powell.[30] According to the director, "I like the repetition of image and sound at the same time, the visual language of the film reinforcing the text. So that in the film you hear Travis say, 'I went to open the door,' at the same time you see him open the door. Narration presents a way of shaping a film."[31] But unlike the voiceovers of hard-boiled detectives used to advance film noir plots, Travis's voiceover is an unreliable form of narration. At the same time that the voiceover is used to draw the audience closer to the character, it also expresses his internal fears and neuroses. At first, the narration is observational, reinforcing what we see in the image while also presenting Travis's perspective. At the same time, the narration reinforces those elements that are significant to the story. Powell explained the device further: "It is a technique that is obsessive in itself simply because it magnifies. We see what we hear, or as we hear it and so it sinks into our minds not once but twice."[32] Or, to reverse Powell's equation, the repetition of details through the visual and acoustic realms reinforces Travis's obsessive nature.

As many critics have pointed out, Schrader's use of voiceover to redirect the audience's understanding of the story comes not just from American film noirs, but also from the films of Robert Bresson. Instead of reinforcing the diegetic details, as Powell asserted, the voiceover fits the model of Scorsese's double articulation of sound. Leighton Grist described its function as follows:

> This visual repetition of the voice-over reflects Bresson's combination of voice-over and shots of the Curé d'Abricourt writing in his diary in *Journal d'un curé de campagne*. It is through such techniques—what

> Schrader, following Susan Sontag, terms "doubling"—that Bresson creates the "disparity" that Schrader considers vital to transcendental style: "his narration does not give the viewer any new information or feelings, but only reiterates what he already knows . . . because the detail is doubled there is an emotional queasiness, a growing suspicion of the seemingly 'realistic' rationale behind the everyday."[33]

This "growing suspicion of the seemingly realistic" is key to understanding the narrative movements of *Taxi Driver* and Travis's underlying motivations. As Patricia Patterson and Manny Farber noted in their review of the film, "*Taxi Driver* is actually a Tale of Two Cities: the old Hollywood and the new Paris of Bresson-Rivette-Godard,"[34] and reading the conventions of classical Hollywood cinema against the ruptures in the film's form that were borrowed from the French New Wave yields interesting insights into its workings.

Aside from its use as a Bressonian motif, the voiceover also makes us aware of Travis's social awkwardness. Painfully uncomfortable in public, Travis regularly retreats into his own internal reveries because, as Powell observed, "He never speaks because he thinks no one will listen to him."[35] Instead, his voiceover forms another story—sometimes reinforcing the images seen, sometimes contradicting them. As the film progresses, however, the voiceover becomes less frequent and occasionally breaks down. At the beginning of the film the voiceover supports the details: we see his world while we hear his opinion of it. Later the focused, contemplative voiceover is replaced with Travis's speculative, doubting thoughts. After his first date with Betsy (Cybill Shepherd), a political campaign worker who catches his fancy, he realizes that he doesn't know her last name and his voiceover interjects, "Betsy what? Oh damn, I forgot to ask her again! I have to remember." What began as diary entries starts to become internal narration, and the viewer weighs the effect of the voiceover against Travis's deteriorating mental state.

One of the main clues that signals Travis's imminent mental breakdown is when the internal narrations shift from his diary entries to more random thoughts. Scorsese explained that Travis's diary "[is] not written at all the way he speaks. He speaks in strange halting ways but his diaries are the diaries of a poet."[36] Yet his internal narration does change through the film and what started as eloquently written observations slowly become self-doubts and platitudes. Mirroring Bresson's Curé d'Abricourt, Travis wonders whether the existential void he feels is stomach cancer, only to reassure himself that "you're only as healthy as you feel." Unlike Bresson's character, Travis's self-doubt is framed not as a religious crisis but rather a personal one. His internal narration becomes more deluded and confused, and after being jilted by Betsy he purchases an arsenal of illegal weapons and practices his aggressive banter in front

FIGURE 17. Travis's (Robert De Niro) internal monologues become unstable and increasingly unreliable in *Taxi Driver* (Martin Scorsese, 1976).

of the mirror. At first the scene is entirely diegetic, yet as he continues it shifts to a voiceover that breaks down and serves as a counterpoint to the images. Leighton Grist described the scene as follows: "As Travis turns his head, in slow-motion, toward the camera, his voice-over resumes: 'Listen you fuckers, you screwheads, here is a man who would not take it anymore, who would not let . . .' The voice-over abruptly stops, then restarts, with Travis repeating his opening words and the scene jump-cutting to a reprise of the shot of his head turning. On one hand, the non-classical editing and camerawork reflects and keeps us 'inside' Travis's perception. On the other, they mark Travis as unhinged."[37] This moment of audiovisual derangement signals a crucial break for Travis, and the voiceover takes on an authoritative tone the second time around, articulating Travis's vigilantism, stating "here is someone who stood up" even though the image reveals him curled in a fetal position on his cot. In a rare moment of calligraphic synchresis, the words of his voiceover—"Here is . . ."—are heard as the film cuts to an image of the same words being written in his diary. As the writing ends in an ellipsis, so too does Travis's internal monologue.

After this point there are only two other moments of voiceover in the film, both based around letters. The first is when Travis writes an anniversary card to his parents, the words of which are heard as Travis's voiceover (see figure 17). But instead of the rambling self-doubt and broken subjectivity of the earlier scenes, the card is filled with banal sentiment and bromides. As he wishes his parents well, Travis explains, "My sensitive work for the government demands the utmost secrecy" and reassures them: "Don't worry about me—one day there will be a knock on the door and it'll be me." The latter line comes off as more of a threat than an assurance because we realize that Travis is clearly not in his right mind. Even though the audience may not notice it immediately, this is the last time that we are granted access to Travis's thought process through voiceover. The rest of the film continues without narration until after Travis's shooting spree to liberate the young prostitute Iris (Jody Foster). Travis survives the film's bloody climax and as an ironic coda we hear Iris's father's voice narrating his thank-you note to Travis while the camera pans over the note tacked to the wall of his apartment.

By structuring the film around Travis's increasingly divergent voiceovers, the film challenges the audience to read the story from a different perspective. As Grist articulated this effect: "Whereas narrative events are linear, narrative logic tends to be associational and symbolic, presenting a subjective structure that invites interpretation."[38] A correlative associational logic can be applied to the use of music in the film. Bernard Herrmann's score music differs greatly in both tone and texture from the diegetic source music used in Scorsese's earlier films, but it too can be read as following a pattern of double articulation. On first listen, one is struck by two aspects of Herrmann's score: its overt presence after the eschewal of score music in Scorsese's prior films, and its similarity to Herrmann's earlier scores, especially those for Alfred Hitchcock's films. Regarding the first aspect, the score is not only present but insistent, very often asserting its presence through the suppression of other sounds. Patricia Patterson and Manny Farber commented that "New York street noise is replaced by a writhing, intense Bernard Herrmann music track, saxophones and a pulled-taffy Muzak sound that almost buries the visuals."[39] And regarding the second, many of the themes don't function simply as underscore because they summon up other films and genres. Charles Michener observed that the film's score functions on two levels: "On the one hand it underscores, in the most old-fashioned manner, the film's every dramatic moment and shift of mood, approaching at times the effect of a sledgehammer. On the other hand, it heightens, by its very qualities as old-fashioned movie score, the dream-like nature of *Taxi Driver*, proclaiming it not a slice of life but an exercise essentially in surrealism, a *movie*."[40]

In keeping with Schrader and Scorsese's modern take on Bressonian voiceover as an insight into Travis's fractured mind, Herrmann's score can also be read as an extension of the central character. Instead of the standard empathetic Hollywood score, Herrmann's music is an "anti-score" that functions dialectically with the audiovisual materials of the film. As a marker of his psychosis, Travis doesn't hear voices; he hears music. The lugubrious saxophone and jazz-influenced "Taxi Driver theme" is heard first over the film's opening and is closely associated with Travis's perception and his nocturnal travels around the city. After a cab emerges from a cloak of subterranean steam the theme plays over its rain-dappled windshield as the lights and colors seem to melt through a step-printed animation effect, ending with a tight close-up on Travis's eyes surveying the denizens of the city. The associational patterns developed in the editing can also be applied to the relationship of the score music to Travis. Three main motifs recur in Herrmann's score: the major chords and saxophone of the romance motif ("They Cannot Touch Her"), the muted minor-key horns of the noir/anxiety motif ("Thank God for the Rain"), and the martial drumming of the military motif ("God's Lonely Man"). If the voiceover is the sound of Travis's ego, then the score functions as an internalized superego whose motifs are drawn from classical Hollywood romance, film noir, and war genres. The changes in music signal the generic lens through which we are supposed to understand Travis's view of his world.

Unlike his voiceover, which tends to rationalize the events occurring around him and often contradicts the visuals, the music hints at Travis's true subjective state and becomes a metaphor for his emotions. Even though it is informed by scoring conventions, the music does not function as a classical score. Instead, changes in music and sound operate as emotional gearshifts, and the viewer learns to listen for these changes to understand their significance. As previously mentioned, there are several moments when the musical score envelops the sound track and literally squeezes out all other sounds. When Travis enters the cab company to apply for a job, not only does he emerge from a cloud of smoke, just like the cab during the title sequence, but he is also accompanied by a classic Herrmann stinger chord that evokes the scene changes from Hitchcock's films. As the scene continues, individual sound effects are introduced—first a ringing phone and rustling papers, then foreground voices, background ambience, and background sound effects—to build the acoustic verisimilitude of the garage as the score fades out. After he leaves, however, the music gradually returns and eventually takes over as his cab prowls the city streets. Only occasional sound effects are heard when the music is playing over the cab sequences to focus the audience's attention on certain details—the splash of puddles, the rhythmic tapping of high heels, or the dialogue of his clients. All these sounds are heard from Travis's point of audition, which places us in the intimate space of the cab, but despite his impassive expressions the shifts in the score tell us how we are supposed to understand Travis's emotional responses.

Modulating dialogue, voiceover, music, and sound effects is the main aesthetic technique that Scorsese uses in the film to direct the audience's understanding of Travis. After the audience learns how to read Travis's internal state through the voiceover and musical score, Scorsese introduced variations in sound use to indicate Travis coming apart. At some points, the score music indicates an internal state that is in contrast with Travis's external actions or dialogue, whereas at other points even his internal narration and the score music are in conflict. "While Travis is speaking of himself," observed Robert Kolker, "the image and music track lead in quite different directions." Even though Travis's voiceover expresses his belief "that someone should become a person like other people," in contrast, "Bernard Herrmann provides a thudding sound almost like a heartbeat."[41] The martial beats of Herrmann's music trump the hackneyed platitudes that Travis intones as his bellicose nature rises to the surface.

Overall Scorsese constructed a system in the film whereby the standard codes of cinema were warped and modulated to mark the changes in Travis's mental state. Both his voiceover and the nondiegetic music signal these shifts, and the sound effects and sound mix of the film also work to ground these changes in diegetic realism. Scorsese's main collaborator in sound was supervising sound effects editor Frank Warner. Warner, a seasoned veteran of the studio system, transitioned to a freelance sound editor in the 1960s and worked closely with directors Arthur Penn, Hal Ashby, and Peter Bogdanovich in the 1970s. *Taxi Driver* was the first film that Warner worked on with Scorsese, but their collaboration would continue with *Raging Bull (1980)* and *The King of Comedy* (1982). In conversations with the director, Warner expressed his main principle guiding the effects work on the picture: "Keep it dirty." As he explained, "You're working in an area of town that is not a happy place. What you would hear from the tenement was not nice—babies crying, radios playing—you just tried not to be pretty or nice with it. There was a lot of water—it was generally wet."[42] The result of the sound mix is that there are sounds that often extend beyond the space of the scene that flesh out the world in which Travis is living. During the scenes in his apartment there is a vast extension of sounds whenever Travis is alone as a reminder of the city life that continues outside his walls.[43] The sounds of children playing, car horns, and traffic filter their way into his apartment, and when he stops to pick up Iris on the street we hear police sirens and radio static coming from distant, unseen sources.

The strategy behind the sound work was to keep the audience aware of these sounds and the fact that their sources could intrude on Travis's world at any time. Moreover, increasing the brittle high end of the sound effects emphasized the sounds of breaking glass and metallic cracks that keep the audience slightly on edge. After Travis buys his arsenal Warner used the scenes of him practicing with the guns to foreshadow the events of the film's violent ending. According to Warner, he decided to "be mean" with the sound effects and make Travis's weapons sound "as mean as you can, because this guy was setting out to do bad things. So make it sharp and hard."[44] The story builds to its climax when Travis decides to assassinate the presidential candidate Charles Palantine

(Leonard Harris). After his voiceovers end, the audience loses access to Travis's inner thoughts and has to speculate about his motivations. When the camera lifts to reveal Travis arriving at the political rally wearing a freshly shaved Mohawk hairdo it triggers our suspicions, but confirmation of his mental state comes during a moment of sound and image disconnect. Even though each of the shots from the campaign rally features realistic sound and image scale matching, there is a moment when this pattern is broken, and the shift in sound perspective occurs at the very moment when Travis becomes unhinged. Travis is seen at the back of the crowd in extreme long shot from a camera position directly behind Palantine, yet we hear the candidate's voice diminished and reverberant from Travis's point of audition. The effect is disorienting and precipitates Travis's advance on the bandstand, with the heightened sound of his jacket's zipper signaling his intention to draw his gun, only to be foiled by Palantine's secret service officers.

This sound reversal also signals a narrative reversal in the scenes that follow. Foiled in his assassination attempt, Travis turns his attention to liberating the teenage prostitute Iris from her pimp, Sport (Harvey Keitel). The apocryphal story is that Scorsese had to desaturate the colors during Travis's violent rampage to secure an R rating for the film, but the desaturated images are themselves counterpointed by hyper-detailed sound effects. As Frank Warner articulated the director's strategy, "Scorsese's only worry was what the gunshots at the end of the picture were going to sound like. That was major. This picture was being made for that moment."[45] Travis's slow-burning aggression, sensed throughout the film, erupts in a climactic display of violence punctuated by Warner's precise sound work. "It's a very important story," explained Warner. "So I went in to make it as violent as I could. I made the shots, seven, ten, eleven sounds combined. Those are the best kind of shots because they're all different. When you saw the shot and the bullet hit, you heard it hit. You heard it hit bone, and you heard it hit blood and the splattering on the wall."[46] More than the graphic display, the sounds of the attack reinforce the primal and visceral emotions at work in the scene. The exaggerated sounds of gunfire and blood during the final sequence render Travis's unraveling, especially because the entire sequence and the scenes leading up to it play out without any score music. "The scene is horrific," explained Leighton Grist, "even more so when, following the massacre's reports and screams, we are left with a silence broken only by the unsettling sounds of dripping blood and Iris's sobs."[47] Only after Travis attempts to kill himself, and we hear the clicks of his empty gun, does the score music return. This time, however, the score overtakes all the diegetic sound effects and the muted horns and martial drums accompany a gliding top shot over the carnage and out into the street, reconnecting Travis to the stunned denizens of the city.

Scorsese's method in *Taxi Driver* was to set up a complex dialectical relationship between the visuals and the diegetic sound effects, nondiegetic score music, and voiceover narration as a guide to the subjectivity of the film's main character. In the coda that follows the massacre, the film returns to a state of normality and a different voiceover is heard when Iris's father's voice narrates his letter sent to Travis thanking him for his actions. The scene that follows plays out without the support of score music or a diminished sense of diegetic sound effects as filtered through Travis's consciousness. Instead, Travis is seen picking up passengers from a hotel, and for the first time in the film the sounds of his cab and the city around him are heard realistically. The rattle of the cab's body, hum of its engine, and crackle of the radio subtend and punctuate the conversation as the cab passes over the rough city streets, all sounds that previously had gone unheard. This acoustic shift signals to the audience that perhaps Travis has been restored to a state of sanity after his violent catharsis, yet this acoustic realism does not last long. When the film cuts to reveal that the passenger is Betsy, Herrmann's saxophone theme returns and accompanies their brief conversation. Just in case the audience is lulled into a false sense of security by the presence of the romance theme, the camera inside the cab after Travis drops Betsy off pans 180° from her standing on the curb to Travis watching her in the rearview mirror. As he moves to adjust the mirror a reverse stinger chord is heard, its unusual timbre and attack reinforcing the fact that both the effect and the music emanate from Travis's mind. As the cab continues past seedy hotels and exploitation movie theaters, the end credits roll and the saxophone theme is replaced by the martial drums and low horns of Travis's "God's Lonely Man" theme. In the end the audience is not left with the reassurance that Travis's psychosis is in the past, and the unsettling return of a filtered soundscape portends further violence as the screen fades to black.

PART THREE

The Dolby Stereo Era (1975–1980)

8

The Sound of Music

Dolby Stereo and Music in the New American Cinema

Upon first listen it seems like very little changed in terms of score music in the 1970s. The masters of wall-to-wall music scoring were still writing into the decade (Bernard Herrmann, Miklós Rózsa, Alfred Newman, Dmitri Tiomkin), and the legacy of composers like Max Steiner and Franz Waxman lived on in their protégés Elmer Bernstein and Jerry Goldsmith. Yet upon deeper inspection several trends began to emerge and the sound track was reconfigured in new ways. One major trend was the wholesale eschewal of the score in films like *Panic in Needle Park* (Jerry Schatzberg, 1971), *The King of Marvin Gardens* (Bob Rafelson, 1972), *Scarecrow* (Jerry Schatzberg, 1973), *The Day of the Jackal* (Fred Zinnemann, 1973), and *Dog Day Afternoon* (Sidney Lumet, 1975). Filmmakers started to question the way that score music guided audiences and conducted their emotional response to films. Instead of the thickly layered orchestral scores of the classical Hollywood era, 1970s filmmakers also began to experiment with minimalist scoring techniques, such as David Shire's solo piano score for *The Conversation* (Francis Ford Coppola, 1974), Stomu Yamash'ta's percussive textures for *Images* (Robert Altman, 1972) and *The Man Who Fell to Earth* (Nicolas Roeg, 1976), and Gil Mellé's electronic score for *The Andromeda Strain* (Robert Wise, 1971). But perhaps the biggest trend was the programmatic use of popular music in lieu of instrumental scores.

Popular Music in 1970s Cinema—The Compilation Score

One of the first films that moved away from the composed score in favor of a compilation score was Mike Nichols's *The Graduate* (1967). Nichols chose to use the music of Simon and Garfunkel to augment the central character Benjamin's odd relationship with his family and his girlfriend's family. Some of the songs used in the film already had been released on prior albums ("The Sounds of Silence," "The Big Bright Green Pleasure Machine," "Scarborough Fair/Canticle," "April Come She Will"), and Simon and Garfunkel were brought in to write the film's theme song, "Mrs. Robinson." The inclusion of these songs targeted a youth audience and the reception of the scenes underscored by the music was determined by the songs' prior cultural meanings. According to James Monaco, "*The Graduate* used music in a special way. Audiences already knew Simon and Garfunkel's songs, and brought with them to the theater attitudes and associations. *Easy Rider* and numerous other films used this same associative technique. Meanwhile, the number of films without any music track to speak of grew. The new attitude toward music seemed to be more narrowly verisimilitudinous. If there was no apparent reason for music in the film, it wouldn't be added later."[1] As Monaco noted, this displayed a trend in American cinema toward an increased representation of the realism of everyday life. Yet the authenticity of musical selections was generally based on a temporal continuity between the songs and the period being represented rather than a spatial realism of music being sourced in the diegesis. The pattern established in films like *Easy Rider* (Dennis Hopper, 1969) and a slew of late 1960s youth-oriented films including *Alice's Restaurant* (Arthur Penn, 1969), *Getting Straight* (Richard Rush, 1970), *R.P.M.* (Stanley Kramer, 1970), and *The Strawberry Statement* (Stuart Hagmann, 1970) was the use of popular music as score,[2] where the songs subtended narrative elements and their preexisting lyrical associations directed audience emotions in the same way as score music. However, a new generation of filmmakers began to take popular music in a different direction in their films.

American Graffiti (George Lucas, 1973)

American Graffiti was the byproduct of George Lucas's reminiscences of his teenage years spent cruising in his hometown of Modesto, California, and the screenplay was worked into shape by Lucas with screenwriters Willard Huyck and Gloria Katz. In contrast to the sterile futurism of his earlier *THX 1138* (1971), *American Graffiti* was designed to be a form of escapist entertainment based on Lucas's personal experiences. To augment his story, the director structured the film around a never-ending stream of pop tunes from the late 1950s and early 1960s strung together in the form of an omnipresent radio broadcast. Huyck and Katz accommodated this when writing the screenplay by making each scene last roughly two to three minutes: approximately the length of a rock 'n' roll song. The entire design of the film was based around the interplay between the rock songs and the narrative events. To capitalize on this interaction, Lucas brought Walter Murch on board early in pre-production to assist in the overall style of the film's sound, and, as Murch expressed it, "George wrote the script . . . with his forty-fives playing in the background."[3] Michael Pye and Lynda Myles described the construction of the screenplay as follows: "From the beginning, the group—[executive producer Gary] Kurtz, Huyck, Katz, and Murch, as well as Lucas—discussed which tune best went where. They opened the film on a giant amber light; as the camera pulls back we realize it is the marking on a radio dial. The structure of the film comes from the radio program, the songs that disc jockey Wolfman Jack plays. Characters take cues from the music. And Wolfman Jack is the unseen center of it all, father figure as much as circus master."[4] While it was clear that the musical component of the sound track was going to play a central structural role in the creation of the film, there was the problem of how this material was going to be presented to an audience. The positioning of the songs firmly within the space of the diegesis brought up the question of how best to source the sounds within the film. To achieve this, Murch, along with producer Kurtz, came up with the concept of "worldizing" the sounds, or creating a strategy for creatively and realistically positioning the music within the diegetic space of the film.

Gary Kurtz was no stranger to this concept after having supervised the sound effects recording for his previous production, *Two-Lane Blacktop* (Monte Hellman, 1971), and Murch had begun to develop his worldizing strategy on *THX 1138*. Although *American Graffiti* was not the first film to feature rock songs as a musical accompaniment, earlier films like *The Graduate* and *Easy Rider* simply substituted the instrumental score music with lyrical pop songs. Murch and Kurtz were striving to create a verisimilitude missing in those prior films. By sourcing the music in the film, as Peter Bogdanovich had done with the country and western songs in *The Last Picture Show* (1971) to evoke West Texas in the 1950s, the presence of the music would reinforce *American Graffiti*'s 1962 setting. According to Murch, "What we did in *Graffiti* was to really make the sound appear to come from the car radios. Partly it was dedication to natural sound, but also it was because *Graffiti*—unlike *Last Picture Show*—is a musical film from the first to the last frame."[5] Murch's use of the expression "musical film" is telling because it hints at an intimate connection between the music and narrative that had previously been reserved for use in musicals. The music was meant not only to accompany the narrative events, but to interact with and influence those events as well. Jeff Smith noted that the rock music in *American Graffiti* served to strengthen the overall narrative structure, to inform the visual design of the film, and to offer an interpretive schema for critics.[6] However, Murch's manipulation of the music in the film fulfilled other functions that did not enter into Smith's analysis, namely the establishment of a spatial connection between the music and its sources as well as the creation of a level of acoustic realism unheard in previous films.

These goals were achieved by taking a two-hour prerecorded tape of Wolfman Jack's radio broadcast and making a new version of the tape by playing it back in different locations and recording it from varying distances. Murch described the procedure as follows:

> George Lucas and I went out into the backyard of a house in Mill Valley. George had a Nagra and a Nagra speaker. I stood about seventy-five feet away from him with a microphone and another Nagra. We turned both Nagras on, and as the tape rolled, he would slowly move the speaker through 180 degrees, and I would move the microphone through 180 degrees in the opposite direction. So sometimes we'd be pointing at each other, and sometimes we'd be pointing at exact opposite directions, seventy-five feet apart. As a result you get a veil of sound that was just recognizable as music because it was so overlaid with secondary and third-level reflections. As you were listening, it would come into focus for a moment, and then it would go out. We recorded the whole two hours twice. So in the end we had three tracks, each two hours long, that we could lay alongside each other; two different atmospheric tracks where the focus level was thrown slightly out of sync and then the perfectly dry, studio-quality radio show.[7]

The three tracks were intermixed to provide different and changing sound perspectives for each scene. Moreover, the strategy allowed Murch to intermix the tracks to simulate both moving sound sources, such as traveling cars, and multiple sources in the same location. This latter strategy was introduced to accommodate the mixing of the film into stereo as Lucas and Murch originally planned. Universal Studios, however, was unwilling and uninterested in paying to have the film mixed in stereo, in part due to the minimal budget assigned to the project, but also due to the studio's growing interest in the Sensurround process.[8]

Murch was forced to abandon stereo as part of his acoustic strategy due to the exigencies of production demands, and he related his aesthetic design to authors Michael Pye and Lynda Myles: "The sound track was theoretically a radio show, with some tracks re-recorded to sound as though they were coming from the cars. As a car drove by, you would have the sound of music driving by. All through the film, people would be swimming in a soup of sound."[9] Even though the concept of worldizing was a very strong attempt to reinforce the diegetic realism through the soundscape, audiences were perhaps not ready to appreciate its effect. According to Pye and Myles:

> In the film as it was originally released, that effect does not quite work; even the giant radio dial in the first shot is not quite enough to make the point. Numbers sometimes echo action or character very patly. In the

> stereo version that Lucas insisted was his condition for allowing a sequel to be made, there is complex crosscutting between speakers at the front and back of the cinema as well as from side to side. The effect is stronger. If you remember, *we lived with the radio on*; music frames and counterpoints the action.[10]

Despite Pye and Myles's misgivings about the monophonic mix for the film, the worldizing effect is quite stirring in the way that it allows the musical sound track to function as both a running commentary on the action of the characters and a strong tool to establish the realism of the diegesis. Although several songs were used to create simple associations between narrative events and lyrical content, Murch was able to elevate the radio broadcast to a higher level by carefully and discretely matching the music to its various diegetic sources. As such the radio broadcast became a structural paradigm onto which all the other narrative material coalesces. Instead of the songs operating as external commentary on diegetic action, the opposite occurs; the radio broadcast becomes the central element of the film, dictating not only the editing choices but also the nature and character of all the image and sound strategies in the film.

The use of the running radio broadcast assists the development of characters and the audiences' understanding of the events in a number of ways. First, the music functions as a narrative aid and helps to cover some of the weaker details in the story's progression. Because the constant shift among the eight central characters causes occasional slips in the temporal progression of the narrative, the omnipresence of the music helps to shore up the many gaps. Second, as reviewer Irene Kahn Atkins pointed out, "*American Graffiti* is a film that is most noteworthy in the visual aspects of its nostalgia, rather than its dialogue which tried too hard to sound the way that American high school students talked in the early Sixties. The music tracks help to overcome some of the deficiencies of the speech tracks."[11] Without the music, the stilted period nostalgia would have interfered with the progression of the narrative.

Oddly, the additional value of the songs does not overload the film with nostalgia; it creates a dialectical tension between the action of the characters, the lyrical content of the songs, and the rigidly realistic sound of the music.[12] If Lucas had simply used the music as an accompanying score to offer wry commentary on the actions, then the film would not have succeeded at establishing the mood that is so central to the story's progression. However, with the rigorous spatialization of the sound of the music, Murch constructed a world where the characters are immersed in music that emanates from myriad sources, but whose sources are never truly shown or localized. This is entirely in keeping with Curt's (Richard Dreyfuss) quest to find Wolfman Jack, the mysterious DJ

FIGURE 18. Near the end of *American Graffiti* (George Lucas, 1973), Curt (Richard Dreyfuss) believes he has discovered the source of Wolfman Jack's radio broadcasts.

who surreptitiously accompanies the characters' lives with his radio show (see figure 18). Not only does the film tell the story of eight teenagers on their last day together, but it is also a story about the search to return an acousmatic voice to its proper body. In this way *American Graffiti* also plays with the question of sound, its source, and its manipulation and reconstruction in recording and reproduction.

In addition to Murch's extreme attention to detail in crafting the film's sound, a large part of the film's success can be credited to Lucas's astute sense of timing and marketing. Although the film itself was not solely responsible for launching a wave of 1950s nostalgia upon its release in August 1973, it did tap into an existing climate. Jeff Smith notes that the film came immediately after a series of well-attended 1950s revival shows promoted by Richard Nader from 1969 through 1973.[13] Lucas understood that the individual songs not only were central to the structure of the film but would also, by drawing on a canon of previously recorded popular songs, form the basis for ancillary marketing of the film's sound track. As James Monaco noted, the film was a success not simply because of its look or the skill of its unknown actors "but because Lucas had the smart sense to mold the film around forty-one rock 'n' roll classics. They were written into the script, and $80,000—more than 10 percent of the budget—went to pay for the music rights. It was worth it. . . . Within a few years, rock 'n' roll would prove to be Hollywood's most financially valuable tool."[14] The rock revolution that Monaco mentions did not see the syntax of film change in any appreciable way, but the benefits of tapping into a youth market through ancillary marketing were exploited by nearly all the major Hollywood studios. Even though *American Graffiti* was not the first film to feature a compilation score, it was the first to consciously utilize the nostalgic value of the songs as a commercial means to promote the film to its chosen market.

This strategy of nostalgia is one that has clearly guided George Lucas's cinematic evolution, from the influence of Flash Gordon movie serials on the *Star Wars* saga to his increased interest in producing fantasy films for younger audiences. But of all the lessons that Lucas learned while working as a young filmmaker, the one that he never forgot was the power of nostalgia in relation to marketing films. Every one of his films as director or producer bears the mark of the lessons learned making *American Graffiti*. However, this mode of nostalgic filmmaking, while commercially proven, often tended to be creatively limited. Several writers have criticized Lucas for inaugurating a nostalgia craze that led to the stylistically bankrupt music films of the late 1970s, like *The Buddy Holly Story* (Steve Rash, 1978) and *Sgt. Pepper's Lonely Hearts Club Band* (Michael Schultz, 1978).[15] As external marketing and cross-promotion of films through sound track releases gained greater efficacy, the cinematic aspects of film construction quickly took a backseat to the financial potential of the sound track tie-in. The massive success of *American Graffiti* led to other hits such as *Saturday Night Fever* (John Badham, 1977) and *The Big Chill* (Lawrence Kasdan, 1983), with nostalgia rapidly dominating the film industry as a highly effective form of marketing. But in the period immediately after *American Graffiti* there were still a number of filmmakers who experimented with the possibility of using prerecorded song scores in a different way. Instead of tapping into nostalgia, filmmakers such as Martin Scorsese, Monte Hellman, and Hal Ashby were interested in the allusive power of popular music and its function in their films.

Hal Ashby—*Harold and Maude* (1971), *Shampoo* (1975), and *Coming Home* (1978)

In particular, Hal Ashby can be seen as a progenitor of an integrated approach to the use of popular music in cinema. Having gained his training as an editor, and after working closely with director Norman Jewison, Ashby received a chance to direct his first film when Jewison tapped him as his replacement on *The Landlord* (1970). For this film Ashby brought in rock guitarist and composer Al Kooper to create instrumental and vocal songs for the film. To capture the mixed race story in the music, Kooper worked with R&B vocalist Lorraine Ellison and gospel/R&B group The Staples Singers on several of the numbers. Unlike the way the lyrical context dictated the interpretation of the montages in *The Graduate* or the comical correspondences between songs and scenes in *American Graffiti*, Ashby often used the music in *The Landlord* in counterpoint to the narrative scenes, constructing a more complicated relationship between image and sound. This carried on in his second film, *Harold and Maude*, where he used songs from British singer and composer Cat Stevens to complement the film's April–December love story. As in *The Graduate*, several of the songs from Stevens had already been released, and he composed two new songs for the film: "Don't Be Shy" and "If You Want to Sing Out, Sing Out."

It was immediately apparent that *Harold and Maude* was a very different film from what audiences expected in 1971. In the introductory scene we see a tight medium shot of the lower half of Harold's (Bud Cort) body. As he moves through a room of ostentatiously baroque furniture the audience hears the direct sound of his footsteps and actions. He then puts on a record, Stevens's "Don't Be Shy," and as the song continues Harold meanders around the room, lighting candles and making preparations just outside the frameline. Before the audience is fully aware of what's happening they see him tie a noose, step into it, and tumble to his "death" as the music abruptly stops. Of course there is no narrative reason why the music should cease playing on the turntable, though the dramatic logic justifies an abrupt cessation of Stevens's song *in medias res.* The interruption of the song has two effects—it points out the material nature of the song and its function within the film (it is literally being played from a record), and its interruption makes the audience aware of the music's nontraditional function as score music. Yet it is perhaps a third effect that makes the film somewhat unique in relation to the early 1970s. With the abrupt end of the song the audience is asked to listen closely to the diegetic sounds and their function. Specifically, Stevens's songs never play through to completion in the film and are heard only in bursts that simultaneously make the audience aware of their presence without carrying on long enough to function as score music.

In addition to challenging audience sensibilities with Harold's penchant for faking suicides, Ashby plays with audience perceptions, especially in relation to offscreen sounds, throughout the film. When Harold is visiting his Uncle Victor (Charles Tyner), an officer in the army, the sounds of unspecified military marching and shouts crescendo as his uncle tries to dissuade him from his relationship with Maude (Ruth Gordon), the climax of his soliloquy coinciding with a round of gunfire. There is an indication that the sounds are synchronous due to their highly reverberant qualities, yet we never see the space of the military camp where his uncle's office is located. In addition, the film features regular use of sound advances to signal the next sequence. A particularly poignant instance occurs when the sounds of a marching band are heard outside the nontraditional meet-cute between Harold and Maude at a funeral. Another element in contrast with the use of prerecorded popular music is the film's reliance on location sound being "sweetened" with added sound effects. Sounds are often added either to cover uneven production sound or to build up to a dramatic climax, such as the military airplane in the distance when Harold's mother (Vivian Pickles) is interviewing his date or the sub-sequent burst of fire sounds associated with Harold's Vietnam-protest-style self-immolation.

Throughout the 1970s Ashby's sound tracks fell into two distinct categories—those that used prerecorded popular music and others that used more stereotypical scores. In the case of the latter, Johnny Mandel composed a military-themed score for *The Last Detail* (1973) that reflected the reality of navy "lifers" Buddusky (Jack Nicholson) and Mulhall (Otis Young) as they escorted Meadows (Randy Quaid), a young kleptomaniac, to a lengthy sentence in Portsmouth Naval Prison for petty theft. *Bound for Glory* (1976), Ashby's biopic of folk musician and activist Woody Guthrie, featured a broad range of familiar American folk tunes adapted by Leonard Rosenman along with fragments of several Guthrie songs, both original recordings and some performed by the film's lead David Carradine, yet the complete versions of the songs were tantalizingly withheld until the film's finale. As Joseph McBride explained: "Fragments of 'This Land Is Your Land' (Woody's most famous and arguably his greatest song) are heard throughout the film, as he hums and composes it in the course of his wanderings. Only at the very end is the song given full-blown treatment, in a moving coda that includes a collage of Guthrie songs by such singers as Odetta, The Weavers, Judy Collins, and Arlo Guthrie. It is then that Woody is finally transmuted into legend."[16] And in *Shampoo* and *Coming Home*, Ashby continued to refine his use of popular music in lieu of conventional scores.

The main function of the music in *Shampoo* is to root the film concretely in its 1968 Los Angeles setting—both in terms of the time the songs were released as well as the mood expressed. Starting with The Beach Boys' 1966 hit "Wouldn't It Be Nice" heard over what seems to be a black screen, the song is immediately interjected with gasps and moans.[17] Only after allowing their eyes to adjust does the audience notice that the ecstatic cries are emanating from a dimly lit room as George (Warren Beatty) and Felicia (Lee Grant) are engaged in lovemaking. Audiences may have been stymied by this beginning since the sounds of orgasmic pleasure were the purview of X-rated films and generally suppressed in mainstream Hollywood offerings. The source of the sounds is confirmed when the phone rings and a series of fumbling sounds are heard as George attempts to answer it. Even though the character voices match the space of the room, the music has no spatial characteristics to indicate that it is diegetic. Instead it takes on a strange meta-diegetic function, hovering indeterminately between diegetic and nondiegetic. While it can be assumed that George and Felicia are listening to the song, the fact that it starts at the beginning of the film, overlaid by the title credits, with no visual source makes it impossible to determine its ontological status. This is significant because the song will return at the end of the film with its associations radically transformed from the beginning.

In contrast to the opening scene, nearly all the rest of the music is diegetically sourced, with the exception of four lyricless musical textures from Paul Simon reappearing in brief fragments. From The Monkees' "I'm a Believer" in

FIGURE 19. The nonstop party sequence from *Shampoo* (Hal Ashby, 1975) juxtaposes the rise of the political right in the 1968 presidential election with the decline of the counterculture.

George's hair salon to Herb Alpert & the Tijuana Brass's "Spanish Flea" at Felicia's house, all the songs are made to sound as if they are emanating from diegetic sources, even if those sources are not revealed. In the final third of the film Ashby created a tour de force of diegetically sourced music when all the protagonists converge on a party with rock music echoing throughout a mansion and its grounds. Upon George and Jackie's (Julie Christie) arrival the familiar yet muffled strains of "Sgt. Pepper's Lonely Hearts Club Band" are heard emanating from inside (see figure 19). As the camera cuts to different locations around the house and grounds, the music takes on spatial characteristics that match the surroundings. In the main room the music is heard directly, distorted due to its aggressive volume and with no reverberation. But when George and Jackie make their way to the tennis courts and bathhouse at the back of the property the sound scale changes, adding more reverberation and rolling off the higher frequencies. Though this may be a relatively simple accomplishment for a seasoned re-recording mixer like Richard Portman, it creates diegetic verisimilitude and works powerfully to anchor the viewer in the time and place of the characters.

Even though the volume and reverberant qualities of the music match their spatial location, the songs themselves are clearly anachronistic for a 1975 audience. As David Ehrenstein pointed out at the time, "The final party sits like a remote fun house—an anti-Disneyland where strobe lights flash and clouds of pot fill the air while the sounds of *Sgt. Pepper* cover everything in a promise of liberation and freedom. But things were never liberated and never free."[18] The songs that follow in rapid succession—Buffalo Springfield's "Mr. Soul," Jefferson Airplane's "Plastic Fantastic Lover," and Jimi Hendrix's "Manic Depression"—solidify the film's setting on the eve of the 1968 presidential election between Richard Nixon, Hubert Humphrey, and third-party candidate George Wallace. Adding to the realism effect, "Mr. Soul" is abruptly stopped by the sound of a needle being dragged off a record and replaced by "Plastic Fantastic Lover." Ashby noted that his main goal in this nonconventional sound track was to set the period. "I've had a lot of criticism from people who would have preferred a conventional score. The music was so much a part of the time. I knew it would bring people back. . . . I'm supplying the images and situations that are allowing the audience to drift. I'm guiding them. It's just another dimension."[19]

Beyond this, however, the music also comments on the vacuity of the participants and the end of the 1960s dream of freedom as perpetuated by popular culture. Each of the musical choices, from groups invariably associated with the 1960s counterculture, carries with it a message of social resistance and the potential for meaningful change. Yet within the context of the film any sense of political activism or even political choice is conspicuously absent. The only campaign advertisements that are seen are for Nixon-Agnew, and these foreshadow the inevitable results of the election. Similarly, the characters show more interest in their personal dalliances than with the democratic process, and despite being set on the day of the 1968 presidential election we never see any of the characters vote. Thus the solipsistic hedonism of the never-ending party taints the musical selections with a bittersweet sentiment, and the potential radical nature of the lyrics is lost amidst the marijuana haze.

Ashby used the music in *Shampoo* to provide a critical commentary on the characters, their actions, and the period, and a few years later he returned with *Coming Home*, a film that also used popular songs to set the period of 1968. Yet in this later film the music is much more programmatic. Instead of a strong counterpoint between music and images, *Coming Home* features a near one-to-one relationship between the narrative scenes and the songs on the sound track. The songs are often sourced diegetically, as in *Shampoo*, but they regularly transition from one time or place to another, where they become nondiegetic. For example, early in the film the character of Billy (Robert Carradine), a mentally unstable veteran, is first introduced with Jefferson Airplane's "White Rabbit" playing in the background. While the music can be taken as diegetic—the scene takes place on the veranda of the military hospital and it is likely that the song could be coming from a radio nearby—the accompanying lyrics of "one pill makes you larger and one pill makes you small, and the ones that mother gives you don't do anything at all" provide an all-too-precise assessment of the pharmaceuticals used to manage Billy's mood swings. A similar effect occurs later when the psych wing of the hospital is shown accompanied by Jimi

Hendrix's "Manic Depression." What's more astonishing is that in this film Ashby shows little consciousness of the critical use of the song three years prior in *Shampoo*.

In part this may be the manifestation of two large changes occurring in American cinema at the time. First, the rock compilation score was shifting from being an anomalous practice to the norm for films that targeted youth audience in the late 1970s. The compilation score usually was tied to a sound track album release that served as a cross-promotional device to advertise the film and to capitalize on its subsequent success. The record companies quickly discovered the advantage of having one of their artists included on a compilation score and subsequent sound track album, and as a result the cultivation of such deals became much more complicated by the late 1970s. Jerome Hellman, the producer of *Coming Home*, explained that the creative rationale for the music "really grew out of an attempt to recall the period emotionally and to implement specific scene material with music that would reinforce the feelings that we wanted the scenes to provoke."[20] But, as he went on to note, "It was first an administrative problem to clear all this music and all these artists."[21] By 1978 it had become substantially more difficult to navigate the intricacies of mechanical and performance rights, and it was only Hellman's close connection with many of the artists on the sound track that made it possible for them to appear: "I remember establishing certain top fees—obviously we couldn't pay then what they might have been able to command in the open market for a single number. I established a 'most-favored-nation' situation, and that's how we approached the superstars. I found that the most helpful element in all my negotiations, though, was that without exception they were one hundred percent behind what the film said, its point of view. They were all lined up with us in their attitude towards the war."[22] Yet, despite their liberal use of music, and unlike *The Graduate*, *Easy Rider*, or *American Graffiti*, neither *Shampoo* nor *Coming Home* ever had affiliated sound track albums.

A second factor affecting the use of popular music in cinema was the development of new audio technologies that augmented the reproduction of sound in cinemas and accommodated frequency and dynamic ranges that were generally only heard at live concerts, where tuned banks of speakers and powerful amplifiers could reproduce music at stirring volumes and in stereophonic separation. Even though there were some earlier experiments in multichannel reproduction in theaters, from the four-track magnetic stereo releases of *Woodstock* (Michael Wadleigh, 1970) and *Gimme Shelter* (Albert Maysles, David Maysles, and Charlotte Zwerin, 1970)[23] to John Mosely's Quintaphonic sound system for the debut of Ken Russell's *Tommy* (1975),[24] the sound system that set the bar for the remainder of the decade and into the 1980s was Dolby Stereo.

Dolby Stereo, Part One: Music

At the turn of the 1970s, two basic regimes of film sound were known to audiences for different purposes: the standard presentation of monophonic optical Academy prints and significantly more expensive multichannel magnetic roadshow prints. The former was the norm for most presentations. It adhered to the standards of optical sound mixing that were determined in the 1930s. The latter was a holdover from the multichannel boom of the 1950s and featured a variety of different configurations of sound magnetically recorded on the film strip to present high-quality multichannel sound for big-budget studio releases.[25] Roadshow pictures, mostly musicals like *Oklahoma!* (Fred Zinnemann, 1955) or *The Sound of Music* (Robert Wise, 1965), regularly used multichannel stereo mixes as part of their presentation to capitalize on their rarity and to sell audiences the uniqueness of the stereophonic experience. Since the 1950s these two platforms essentially limited high-quality stereophonic sound to big-ticket roadshow presentations, leaving most theaters with optical sound reproduction that sounded scarcely better than it did in the 1930s. There were a large number of predecessors to Dolby Stereo that, for one reason or another, never achieved an industry-wide level of acceptance. Thus Dolby Stereo, which delivered high-quality stereophonic sound in the same space as the standard 35mm monophonic optical sound track, marked a major change in the basic nature of film sound.[26]

In May 1965, Ray Dolby founded Dolby Laboratories to devise a workable solution to the problem of tape noise. The lab's engineers determined that most of the noise on magnetic tape occurred in the frequencies above 3000 Hz. They then constructed a system of filters to break down the frequency range of the recording into four bands with each one channeled into its own compressor to temporarily reduce the dynamic range. The decoding process for playback was essentially the opposite. The signal was again split into four bands and each band received an amount of dynamic range expansion commensurate with the compression in the recording process. In operation the Dolby noise reduction system had two distinct effects: the overall dynamic range of the system's output was greater than the input signal, and unwanted tape noise was eliminated. Thus, this system, referred to as Dolby A-type noise reduction, allowed for the recording and reproduction of source material without the addition of tape hiss even after multiple recorded generations.

But the true genius of Dolby Laboratories was in the way this new technology was marketed and distributed. Instead of manufacturing recording hardware and competing with established companies, Dolby chose to license the noise reduction system on a nonexclusive basis, making it available to all manufacturers. By doing so the firm rapidly ensconced itself in the post-production audio world where the build-up of noise in multichannel audio recording and mixing had been a problem for decades. Dolby's rapid acceptance in music recording and reproduction led to it becoming a household name by the early 1970s, but its impact on film sound was nominal until director Stanley Kubrick approached Dolby Laboratories to use Dolby A-type noise reduction on all the post-production audio for *A Clockwork Orange* (1971). Kubrick used the Dolby noise reduction system to facilitate the extensive mixing of the picture in the re-recording phase, where as many as five generations of magnetic material were overlaid without appreciable noise accumulation.[27] But because there was no system in place to reduce noise in theatrical presentations, and despite the care and attention given to the re-recording of the sound material, the film was eventually released in December 1971 with a standard monophonic optical track. In response to Kubrick's use of Dolby A-type in post-production, however, Dolby Laboratories actively began exploring the idea of applying noise reduction to release prints.[28]

Noting that optical sound on film, like magnetic tape, is an analog medium, the engineers at Dolby Laboratories discovered that 35 mm optical sound tracks could also benefit from the application of noise reduction circuitry. Starting with *Callan* (Don Sharp, 1974), *Steppenwolf* (Fred Haines, 1974), and *Stardust* (Michael Apted, 1975), there were nearly a dozen films made in 1974 and 1975 that used Dolby A-type noise reduction on their monophonic optical release prints.[29] As an ancillary benefit to the company, listening tests demonstrated a compatibility of prints with A-type encoded optical sound tracks when played back in theaters without a Dolby decoder.[30] Even though Dolby-encoded film was reproduced best in a theater equipped with a Dolby decoder, the existing equalization settings for optical sound attenuated the boosted treble without a loss in fidelity. If anything, a Dolby A-type encoded film played without decoding sounded brighter and clearer than a non–Dolby encoded monophonic print. Therefore, Dolby Laboratories believed that production companies and distributors would be more willing to release a Dolby A-type encoded sound track to theaters without encoding equipment if they could reassure the exhibitors that Dolby-encoded prints sounded better even without decoding equipment.

Whereas Dolby Laboratories were looking for multiple uses for their noise reduction technology, stereophonic cinema had never been an interest until an outside researcher presented them with a fully functional stereo variable optical track. Ron Uhlig, an engineer at Eastman Kodak, was exploring the possibility of using Dolby noise reduction on split-channel 16 mm optical tracks to provide dual language choices for educational films as a way to compete with the burgeoning videotape industry.[31] When Uhlig approached Dolby Laboratories with his idea, the company saw a much broader application for the technology. Working in conjunction with Uhlig, Dolby Laboratories designed a 35mm stereo variable-area optical encoding and decoding system known as Dolby SVA (stereo variable area) or Dolby Stereo.[32] Dolby A-type noise reduction was used to restore the fidelity lost due to the reduction in track width created by splitting the monophonic optical track into stereo. Most importantly, because the sound track had two variable-area components, which was nearly identical to the design of monophonic optical tracks, it could be played on any monophonic sound system without the loss of acoustic information, thereby ensuring its backward compatibility.

At the heart of the Dolby Stereo system was the CP100 Cinema Processor which decoded the two-channel sound track into three behind-the-screen channels—left, right, and a derived center channel—to provide a high-quality, multichannel sound track from an optical source. The CP100 decoded a two-channel signal from a stereo optical track by running it through Dolby A-type noise reduction unit and an additional matrix processor to derive multiple channels of sound information from the two stereo tracks.[33] The three elements worked together to produce a virtually distortion-free, low-noise, high-signal output that was played back in Left-Center-Right and eventual Surround configurations. This system was adopted as an industry standard, ISO 2969, for multichannel optical presentations in the 1980s and became the dominant format until the introduction of DTS digital sound in 1992 and Dolby Digital in 1993.[34]

Satisfying many of the needs of the film industry, especially increased sound quality and backward compatibility, Dolby Stereo made its commercial debut in 1975 with the release of two rock 'n' roll–based films: Ken Russell's *Tommy*[35] and *Lisztomania*.[36] Both films—the former based on The Who's 1969 "rock opera" of the same name and the latter featuring Who vocalist Roger Daltrey in a revisionist rock biopic of Hungarian composer Franz Liszt—capitalized on Dolby Stereo to replicate the volume and spatial breadth of live rock performances. The system's derived center channel served to eliminate the "hole-in-the-middle" phenomenon present with two stereo speakers, and the three channels provided adequate coverage on even the largest screens. The next Dolby Stereo film, *A Star Is Born* (Frank Pierson, 1976), also included a fourth surround channel that was encoded onto the two optical tracks and then decoded upon playback.[37] In late 1975, the prior success of the system convinced producer Jon Peters to release the film in Dolby Stereo, but Peters had an additional request, that the film include a surround channel as well as three behind-the-screen channels. According to Dolby Laboratories vice president Ioan Allen: "Initially we were just interested in stereo behind the screen. . . . And, actually, it was a fluke that I was lobbied by *A Star Is Born*, the Streisand film. And they said, yes, we will release this film in stereo optical, but only if you can put a surround on it. And I really wanted that title because of the music in it. So we kludged together the first surround concept in somewhat of a rush."[38] The "kludged together" system, still called Dolby Stereo, became the basis for the four-channel sound system that made Dolby an industry standard by decade's end. Most importantly, its use on several prominent rock films demonstrated to filmmakers and audiences alike the added value of using the Dolby Stereo system for the reproduction of music in cinema.

Dolby Laboratories used a Sansui QS Matrix—a remnant of the quadraphonic music boom of the early 1970s—which utilized phase change relationships to mix Left, Center, Right, and new surround channel information onto two Dolby noise reduction–encoded optical tracks.[39] This allowed for a full four channels of sound to be encoded onto the two-track optical sound track, obviating the need for the more expensive four-channel 35mm magnetic format. Perhaps it was Dolby's prior involvement in the music industry, or the adaptation of the quadraphonic technology for the hardware, but nearly all of the first forays into Dolby Stereo featured music in relation to their screen stories. The representational strategies of Dolby noise reduction system in the music industry rapidly attached themselves to Dolby Stereo without much attention paid to the longer history of stereophony and film, and the result was a kludged-together aesthetic that didn't always accommodate the needs of narrative cinema.

When presenting music, especially music recorded in stereo, the device functioned spectacularly. The noise reduction circuitry expanded the dynamic range while avoiding the distortion associated with overloaded optical tracks. As well, the decoding of the center channel provided the spatial fill needed to prevent the "hole-in-the-middle" effect from widely spaced left and right speakers. But it was the addition of the surround channel that made Dolby Stereo unique as a stereophonic cinema sound system. Because the surround channel was encoded onto the sound track by using phase relationships between the left and right channels, decoding upon playback had an unusual effect.[40] Unlike monophonic dialogue or sound effects, stereophonic music has complex phase relationships that caused the music to be especially active in the Dolby Stereo surround channel. Dolby's choice of testing the surround system on musical films meant that a main criterion in designing the system was ensuring that the musical passages sounded their best. Decoding the surround channel meant that the spatial artifacts of stereophonic music were sent to the surround signal, and the effect was a constant field of sound that surrounded the audience from all sides, similar to either the experience of attending a live musical performance or listening to a quadraphonic recording. Because the surround channel was very active during stereophonic music reproduction, the system rapidly gained favor among filmmakers as well as audiences. This was crucial to a film like *A Star Is Born*, which wanted to immerse the audience in concert ambience for the majority of the film.

When dialogue and sound effects were recorded in Dolby Stereo, however, the system did not function quite as well. Certain frequencies and moving sounds would confuse the Dolby Stereo matrix, which then would send these erroneous sounds into the surround speakers. The result was a "breathing" or "pumping" effect where the sound would swing from the front channels to the surrounds due to either transient sounds with odd phase relationships, such as gunshots or sibilant crashes, or dialogue with hard plosive attacks. To combat these systemic artifacts, Dolby Laboratories required that their consultants oversee the mixing of all Dolby Stereo films. These demands meant that post-production mixing practices had to be standardized to avoid problems in the sound presentation during exhibition. For example, dialogue had to be mixed into the central channel both to ensure comprehension and to avoid phasing problems. Effects could be positioned anywhere in the left-to-right space of the screen, but moving sounds would be carefully monitored to ensure that their motion in the mix matched the motion on screen.[41] Music, however, provided few problems because it was rarely anchored to an onscreen image. But when it was, care was given to make sure that sounds did not drift in the stereoscape.

While music and effects were occasionally sent to the surround speakers, dialogue was strictly avoided. As a byproduct of the surround channel matrix, any playback in mono resulted in the loss of all the surround information. Therefore, to ensure compatibility with monophonic optical projectors, Dolby Laboratories insisted that any information exclusive to the surround channels would have to be expendable. "Surround elements" could be used in the rear speakers as long as they contained no narratively significant information like dialogue.[42] Conversely, all dialogue was channeled to the center speaker to ensure intelligibility. Much like its 1950s magnetic predecessors, Dolby Stereo allowed the sound track to expand into the three-dimensional space of the theater, but only after fixing the voice to the plane of the motion picture screen. Often this led to static sound tracks where music and certain sound effects (like car drive-offs or plane fly-bys) burst free of the plane of the screen, while dialogue and sound remained within the space of the diegesis, locked behind the plane of the image.

As a secondary corollary of the system's design, it was impossible to add divergence to the sound track. Divergence refers to a mixing practice developed in relation to the multichannel magnetic sound tracks of the late 1950s and 1960s. With the use of discrete channel magnetic systems, there was a need for sonic reinforcement to ensure the proper spatial placement of sounds in a multiple-speaker stereophonic field. Audio engineers and mixers were concerned that the spread of the stereophonic field, especially in relation to the widescreen formats available, meant that different seating positions in a theater would result in radically different listening experiences. In particular, due to the Haas effect,[43] individuals sitting on the left or right sides of a theater would have a need for "sound reinforcement" to ensure that they would perceive the sounds as coming from their intended screen location.[44] This meant that if a sound was panned to the center of the stereophonic spread to match its source's screen position, a lesser amount of that sound would be fed into the speakers to the left and right.[45] As a result, sound effects or dialogue could be positioned exactly in conjunction with the screen location of their source, and the reinforcement of the sound through divergence would ensure that anyone in the audience would hear the sounds as coming from the same location.

Yet with the electronics behind Dolby Stereo, divergence was impossible. Because the Dolby Stereo system utilized matrixing to derive both center and surround channel information instead of discrete audio tracks, any reinforcement introduced through divergence mixing in post-production would be eliminated in the decoding process.[46] This meant that traveling sounds, or sounds in the side channels, were much more noticeable than in a magnetic print with divergence. Therefore effects tended either to remain centered on the screen or to drift off into a side channel and the surrounds. In contrast, music reproduced spectacularly and early films in Dolby Stereo were often praised for the sound of their musical sound tracks rather than for any sense of diegetic realism made possible by multichannel sound. Music in Dolby Stereo became an event, and the attraction of the Dolby system was rooted in the experiential nature of listening. Dolby Stereo could replicate the conditions of a live music concert, and the early films that used the technology exploited this effect. As far as the musical use of Dolby Stereo to replicate a live event, no film better encapsulates the experience than Martin Scorsese's *The Last Waltz*, a documentary of The Band's final concert.

The Last Waltz (Martin Scorsese, 1978)

"This Movie Should Be Played Loud!" reads the opening title card of *The Last Waltz*, a film shot at The Band's farewell concert at the Winterland Ballroom in San Francisco on Thanksgiving, 25 November 1976, but not completed until April 1978. In the interim Martin Scorsese released his big-budget tribute to the American musical genre, *New York, New York* (1977), and *The Last Waltz* represented his return to documentaries. Unlike most music documentaries that sought to capture either a particular performance (*Monterey Pop* [D. A. Pennebaker, 1968], *Woodstock*) or a behind-the-scenes view of a performer (*Lonely Boy* [Wolf Koenig, 1962], *Dont Look Back* [D. A. Pennebaker, 1967]),

The Last Waltz wove together performances from The Band and their stellar line-up of guests with interviews with the group almost a year after the concert event. The result is a retrospective view that oscillates between the electric performances and the much more sober analysis of The Band's history and their retirement from touring.[47] But above all, the music in the film is presented as the central attraction for cinemagoers, and Dolby Stereo was used to maximize the experience.

Consequently, the film was planned as a document of their last performance but also as an attempt to use cinema as a surrogate for the live experience. Several previous films attempted to capture the experience of a live concert performance, but most relied on a direct cinema aesthetic of positioning multiple 16 mm cameras throughout the auditorium in the hope of capturing essential moments in the performance. Yet *The Last Waltz* was very different in its planning; according to James Monaco,

> Unlike other concert films, *The Last Waltz* was shot in standard-gauge 35 mm, which gives it a visual presence that is unusual in this genre. Scorsese carefully choreographed his eight cameramen (who included Laszlo Kovacs, Vilmos Zsigmond, and Michael Chapman), referring to a 300-page shooting script which listed all lyrics and chord changes. The soundtrack was mixed down to four Dolbyized tracks from an unprecedented twenty-four-track location "take." ("The longest mix in history," according to publicists.)[48]

These factors made the film radically different from other rock documentaries, even if they also used multitrack recording technologies, because Scorsese had planned out the shooting agenda before the show. Similar to *Woodstock* (Scorsese was one of the film's editors) and *Gimme Shelter*, the structure of *The Last Waltz* was determined partly in the editorial process, but the decision to present the film in Dolby Stereo also had an effect on the film's overall aesthetic.

In particular, the expanded dynamic range and multichannel spatial orientation of the system allowed Scorsese and his sound team to remix the music to match the images. Similar to Robert Altman's practice of "un-mixing" his production recordings by using an eight-channel multitrack recorder on set, re-recording mixers Steve Maslow and Bill Varney were able to take the multitrack music recordings and emphasize particular instruments or vocalists according to their presence onscreen. This had already been done in earlier multitrack musical films, but the fact that the film was being released in Dolby Stereo encouraged the filmmakers to experiment with creating an optimal auditory as well as spectatorial experience. In the process of mixing the music, concessions had to be made to match the location of particular musical elements to their screen location rather than their ideal location in the musical mix.

As an example, if the image shows a wider shot of the stage, the spatialization of the music is spread across the three speakers behind the screen, roughly matching the position of the individual musicians as seen from left to right. In addition, the crowd sounds and the ambient reverberation of the hall are mixed to the surround channel to enhance the concert effect. But more than replicating the experience of being at the concert, the editors and sound team created a new experience for the viewer, as if you are inside the musical performance itself. This was accomplished by favoring certain instruments in the mix whenever the camera isolates a musician, especially during solos. This practice is established early in the film with cutaways to Rick Danko's bass, Garth Hudson's organ, and Robbie Robertson's guitar during the song "Don't Do It" that are matched by an increased presence of each instrument in the musical mix. Occasionally choices of spatialization are also used to indicate the location of a musician on the stage. When Joni Mitchell joins Neil Young for back-up vocals on "Helpless," she is seen silhouetted backstage left and her vocals are panned to the left speaker in the mix. In contrast, when Dr. John performs "Such a Night," his vocals are sent to the center speaker even though the stereo miking of his piano is panned to the left and right speakers.

In this way *The Last Waltz* functions as a curious hybrid of a film. At the same time that they tried to replicate the concert experience for cinematic audiences, the filmmakers had to make choices about which preexisting codes of representation they were going to follow. Unlike most of the previous rock documentaries that presented the performances from a privileged position in the crowd, usually from up close in front of the stage or a more distant long shot at the center of the auditorium to encompass the entire proscenium, Scorsese's cameramen were positioned throughout the performative space and on the stage with the artists. This opened up the sound team to deploy sounds across Dolby Stereo's 360° spatialization and to reconceptualize the relationship between sound and image. With synchronization providing the anchor between the music and the onscreen images, other sound and image relationships could be expanded. During Joni Mitchell's performance of "Coyote" the film cuts to a camera in the crowd that slowly zooms back from a medium close-up to medium long shot, and as it does a slight reverberation effect is added to Mitchell's vocals to emulate the distance change (see figure 20). In Garth Hudson's featured number, "Genetic Method/Chest Fever," the swirls of his organ solo are panned left and right to reproduce the psychedelic experience of the song despite his image staying firmly on the center of the screen.

Curiously, the hybrid aesthetic techniques present in the film ultimately are grounded in an attempt to provide a sense of realism associated with the live concert. Dolby Stereo's expanded dynamic range, low noise, and spatial characteristics were ideal for replicating the sound of the concert. Yet in all the

FIGURE 20. Dolby Stereo technology is used to render the space of the auditorium in *The Last Waltz* (Martin Scorsese, 1978) when the camera zooms back from Joni Mitchell and the reverberation of her voice increases in the surround channel.

documentary interview sequences, the dialogue as well as the related sound effects and ambience were sent to the center speaker as a monophonic track. As J. P. Telotte described it:

> With the documentary approach came an obvious gain in concentration on the musical or expressive elements—the genre's central attraction. Even if the concert film could not quite transport its audience to the event which it purported to record, by recreating some measure of the original musical experience or by even improving it through the use of Dolby sound enhancement, carefully controlled acoustics, and multitrack recordings, it could hope to vicariously furnish an element of the immediacy and vitality of the performances caught on film.[49]

More than re-creating the concert experience, Scorsese sought to improve it by constructing a film that used the visual aesthetics of fiction cinema by adhering to the rules of continuity editing and shot/reverse shot. Importantly, he also used multitrack sound recording and the Dolby Stereo reproduction system to create a form of acoustic continuity where the overall musical mix matched the images even if that meant overriding the standard conventions of live music mixing. While this created an idealized audiovisual perspective for the cinematic viewer, it did so at a cost: the extirpation of the concert audience.

Unlike *Woodstock*, which ecumenically balanced its presentation of the live performances with images and sounds of the audience members to represent the experience of the event and the sense of community it fostered, *The Last*

Waltz developed a new aesthetic based on the systematic exclusion of the audience. Apart from a few brief images of fans cueing around the block to enter the auditorium and a raking pan of the crowd at the beginning of the film, images of the audience are avoided throughout. Moreover, the sound of the audience is surreptitiously excised from nearly all the musical performances and only allowed to creep back in during the interludes to reinforce the illusion of the live event. What Scorsese did in *The Last Waltz* was not to create a record of the event so much as to provide a reconstruction. As Terrence Rafferty observed: "The calculated *mise en scène* tells us that the values celebrated by *Woodstock* no longer bind the performers to the audience, but only bind the performers to each other. The group identity is all on stage now, it's The Band and their guests, now no longer an image of community so much as a community of images, perpetuating on stage what recently existed on a much larger scale beyond it. What *The Last Waltz* is celebrating, after all, is the end of The Band's 'live' relationship to the audience."[50] As a result the film foregrounded the value of the reproduction over the original experience and constructed an aesthetic template for future representations of live music in cinema.

The irony behind the numerous claims that *The Last Waltz* is the best rock concert film is that on the surface the film flaunts its links to realism—the documentary aesthetics, its synchronous sound recordings, and occasional mistakes left to add to the patina of liveness—while at the same time it posits the mediated reproduced experience as being more authentic than the original performance. Not only is there a sense of loss based around The Band ceasing to operate as a live performing unit after sixteen years on the road, there is also a sense of loss surrounding the waning affect of the communal live concert experience. "Ultimately, as *The Last Waltz* rearticulates the modernist nostalgia inscribed within The Band's music," commented Leighton Grist, "so the latter, and its creation, become objects of nostalgia, with the film becoming a lament for that which has, tacitly, been lost twice over."[51] Bound up with this double loss is a commercial imperative based around the synergizing of commodities. Concurrent with the release of the film in April 1978 was the commercial release of a three-disk sound track album, which eventually rose to #16 on the *Billboard* charts and has remained a steady catalogue seller. The music on the album, however, is not the actual mix heard in the film but instead features a more traditional mix that privileges the aesthetics of a live concert recording over the image-determined sound mix from the film. Despite its separate sound aesthetics, the sound track album was complicit in selling a nostalgic experience to audiences—not the actual concert itself, but the cinematic representation of it that was known to most audience members.

As a model for how multitrack sound recordings and Dolby Stereo can be used to reproduce a live musical experience, *The Last Waltz* stands as the exemplar. It is very clearly one of the best sounding films from the 1970s and its influence can be heard in the numerous concert films that followed in its wake. Yet it also represents an empirical break with earlier music documentaries and reinforces a primacy of the cinematic representation of music over the original musical event. "Essentially, Scorsese has fashioned a hermetic or self-enclosed world," noted Telotte, "one admittedly far removed from our ordinary environment."[52] *The Last Waltz* was a harbinger of the radical changes to come in how music was used in cinema and the relationship between the film and music industries. As David Bartholomew commented, "*The Last Waltz* was released at just the right time, before the spate of dumb, prepackaged music shows, geared around potential best-seller sound-track albums, like *FM* and *Thank God It's Friday*, and the summer extravaganzas, *Grease* and *Sgt. Pepper's Lonely Hearts Club Band*."[53] The film stands at a turning point between the use of music in film as a narrative tool and the growing commercialization of sound track recordings with the direct marketing of songs through their inclusion in films in the wake of Dolby Stereo's acceptance.

Merchandizing Music and Synergizing the Sound Track

Perhaps the biggest effect of Dolby Stereo in relation to music is that it not only convinced theater owners of the value of enhanced cinema sound but also solidified the cross-marketing potential for music companies and publishers to feature their songs in films. The best example is the case of *Saturday Night Fever* and Robert Stigwood's marketing of the film and record releases simultaneously. Stigwood, a long-time veteran of the music industry, learned the value of cross-promotional marketing when his management company, the Robert Stigwood Organization (RSO), struck a financing deal with Polygram Records shortly after RSO's incorporation in 1970. In addition to buying the British rights to several American plays and subsequently releasing their sound track albums through his RSO Records imprint, Stigwood bought the theatrical and movie rights to Decca Records' unproduced "rock opera" *Jesus Christ Superstar*.[54] Stigwood's stage productions of the show rapidly netted a total of $62 million in ticket sales, and Norman Jewison's film adaptation took in another $13 million. Adding the sales figures for the original album to the sales of the film sound track album, associated singles, and additional merchandising, R. Serge Denisoff and William D. Romanowski estimated that the entire franchise brought in more than $125 million by 1975.[55]

This synergistic marketing campaign convinced Stigwood that the careful coordination of trends in popular cinema and popular music could generate massive profits. After the relative failure of his second foray into film production with Ken Russell's psychotropic visualization of The Who's *Tommy* in 1975,

Stigwood was quick to recognize the value of the rising disco trend when he purchased the movie rights to Nik Cohn's nonfiction essay "Tribal Rites of the New Saturday Night," which appeared in *New York* magazine in June 1976. Initially Stigwood chose John Avildsen (*Joe* [1970], *Rocky* [1976]) to helm the picture, but when Avildsen's script, which emphasized the dance contest over the love story, was characterized as "Dancing Rocky," the more tractable John Badham (*Bingo Long's Travelling All-Stars and Motor Kings* [1976]) was brought in to direct.[56] Not only did Badham prove more compliant in foregrounding both the love story and the music in the film, but he also wanted the use of music to be in line with earlier examples from the 1970s. "I wanted *Saturday Night Fever* to be filled with the raw impressionistic, gut feelings of *Mean Streets*," said Badham, "where the music enveloped everything and the film was a non-stop dance."[57] While there is no argument that the music in *Saturday Night Fever* does envelop everything else, including four songs from the Bee Gees that went on to dominate the American charts during the first six months of 1978, its programmatic popular music score was a far cry from Scorsese's dialectical use of music in *Mean Streets*.

Instead of the vertical relationship between the songs in *Mean Streets* and the narrative scenes they were set against, the logic behind music use in *Saturday Night Fever* was based on proprietary economics. After their chart-topping 1971 single "How Can You Mend a Broken Heart," the Bee Gees (managed by Robert Stigwood since the 1960s and on his RSO Records imprint since 1973) labored through the early 1970s without much success. Adjusting their sound to match the rising dance trend, they reached the top of the charts again with "Jive Talking" off their 1975 album *Main Course* before fully embracing the disco formula on "You Should Be Dancing" in 1976. Even though each of the songs remained at number one only briefly, Stigwood recognized the cross-promotional potential of coordinating record, film, and sound track releases in order to maximize profits. Striking a distribution deal with Paramount Pictures for 45 percent of *Saturday Night Fever's* theatrical gross, Stigwood retained the rights to the sound track album and released it on his own imprint through Polygram Records.[58]

Because Stigwood was taking a cut of the theatrical box office as producer of the film, receiving profits from the record sales of the sound track album, and earning revenue from Bee Gees songs, both as their label boss and their manager, it was in his best interest to use all these elements to promote each other. During the fall of 1977, Stigwood released the Bee Gees' single "How Deep Is Your Love," which had a slow build-up from its entry into the charts during the week of 24 September until it reached the number-one spot on 24 December.[59] Without a new Bee Gees album to promote, the single featured the caption "From the Paramount/Robert Stigwood Motion Picture *Saturday Night*

Fever" to provide exposure for the forthcoming film and its sound track album. Coincident with the release of "How Deep Is Your Love," RSO produced a thirty-second theatrical trailer for *Saturday Night Fever* and in November released the double-album sound track to retailers.[60] By the time the film was released on 16 December, the sound track album had sold 850,000 copies and continued to sell an additional 750,000 copies over the Christmas holidays.[61] RSO extended its marketing campaign by staggering releases of new singles from the album throughout the first half of 1978. As a result, the sound track album sold over 25 million copies worldwide, bringing in an estimated $285 million, generated four consecutive number-one singles, and helped drive the film, budgeted at $4.5 million, to a box office total of $74.1 million.[62]

Saturday Night Fever set the bar for the profit potential involved in the careful cultivation and marketing of sound track albums and related singles with films. Stigwood continued to link artist management, film production, and sound track recordings to greater (*Grease* [Randal Kleiser, 1978]) and lesser (*Sgt. Pepper's Lonely Hearts Club Band*) success over the remainder of the decade. As James Scale pointed out: "*Saturday Night Fever* not only proved there is a gigantic market for rock music films: it showed the same audience is accustomed to the high-quality sound of costly home stereo systems. New sound recording systems such as Dolby and dbx are now considered a prerequisite to any ambitious musical, since a new sound-conscious audience wants a feast for the ears as well as for the eyes."[63]

In the wake of *Saturday Night Fever*'s success, fully half of the films released in Dolby Stereo in 1978 were music-oriented. These included rock concert films (*The Grateful Dead* [Leon Gast, 1977], *The Last Waltz*), rock-themed musicals (*Grease*, *Sgt. Pepper*, *The Wiz* [Sidney Lumet, 1978]), or films that featured rock compilation scores or scores composed by rock artists (*Big Wednesday* [John Milius, 1978], *FM* [John A. Alonso, 1978], *The Shout* [Jerzy Skolimowski, 1978], and *American Graffiti*, re-released in 1978 in multichannel stereo).[64] The initial advertisements, preview posters, and trailer for *Saturday Night Fever* did not feature the Dolby logo, but it was included in the second round of advertisements and promotion for the PG re-release.

In the minds of cinema audiences, the first wave of Dolby Stereo films was firmly associated with music and its use in film. Moreover, after ignoring the value of the sound track album for decades, filmmakers started to consider the use of compilation scores as a marketing tool. In conjunction with this, a new role emerged in the form of the music supervisor as the liaison between the film producers, the record labels, the publishing houses, and the musical artists. The corporate synergy between the film and music industries became the norm by the 1980s when "film companies saw what it could do for their films in terms of promotion to have songs you could put on a sound track album."

According to music supervisor Becky Shargo, "You got all this air play, all this free advertising. And in turn the record companies saw how the film company's marketing campaign could help promote their record."[65] As a result, Dolby Stereo received its first boost as part of the growing use of popular music in cinema, and its part in the crossover promotion between the film and music industries helped to establish Dolby sound systems in theaters across American and the globe. At the same time that the Dolby name was extending its association with music and music-based films, it was also making inroads into a much-neglected genre, science fiction, as the premiere platform for the new sound aesthetics in the 1970s.

9

The Sound of Spectacle

Dolby Stereo and the New Classicism

With the rise of Dolby Stereo technology in the late 1970s, the American film industry was forced to reconsider how to integrate multichannel sound into established patterns of narrative representation. Although there were a few experiments with older forms of magnetic multichannel sound systems, the rapid acceptance of Dolby Stereo meant that new standards emerged regarding the positioning of voices and spatialization of effects that conflicted with prior uses. Concurrent with the development of these new approaches to recording and mixing film sound was a move toward the specialized construction of sound effects. Sound effects editors and Foley artists started to receive official recognition from the industry, while sound effects construction followed two divergent paths: custom-made sound effects and sound design. The former broke free from the use of prerecorded stock libraries in lieu of the creation of original effects, whereas the latter presupposed a single "sound mind" guiding the overall sound of the film from production through post-production. Both represented a loosening of the strict labor divisions of the sound editing and mixing unions, but whereas the former marked a realignment of creative roles within a hierarchy, the latter became possible through the use of "freelance" sound effects creators and the further fragmentation of roles in post-production sound. Perhaps the best example of these divergent strategies and their effects came with the extensive sound work done on William Friedkin's *The Exorcist.*

The Exorcist (William Friedkin, 1973)

In a 1975 interview, Friedkin was quoted as saying, "There wouldn't be a *Mean Streets* without an *Exorcist*. The studios need box office hits to take chances with the smaller, more personal films."[1] The remark is telling in underscoring *The Exorcist*'s role in introducing a new approach to the economics and marketing of Hollywood filmmaking: a Faustian bargain where the smaller, more personal films made within the studio system relied on the large blockbusters as the source for their funding. Released on 26 December 1973, *The Exorcist* was Hollywood's latest mega-hit (the first since *The Godfather* [Francis Ford Coppola, 1972]), one that reflected an extremely calculated approach to provoking audience responses. The subject matter, drawn from William Blatty's best-selling novel, guaranteed a large audience for the film, yet Friedkin wanted to make it more than just a simple adaptation. His Oscar-winning success with *The French Connection* (1971) gave the director both funding for the adaptation and a large measure of freedom in the production and post-production processes, and it was his desire to shock and terrify his audience by any means at his disposal. Along with a number of unscrupulous methods for eliciting the "proper" reaction from his actors—including submitting them to torturous stunt work, firing guns on set, and even slapping an actor to evoke an emotional response—Friedkin also experimented with a variety of special effects, makeup techniques, and a highly expressive sound track. Although the latter element is under investigation here, it is important to note that it represents only one element that Friedkin utilized to manipulate audience reaction. In many ways *The Exorcist* can be evaluated through its formal attempts to stimulate the audience directly with myriad shocks and surprises. Most importantly, Friedkin and his sound team utilized experimental sound techniques to further these goals, and as a result the film was honored for its accomplishments with an Academy Award for Best Sound.[2]

Very often the result of these acoustic experiments was the pure physical stimulation of his audience. Friedkin claimed that, like Hitchcock, he was attempting to manipulate their emotional responses; yet unlike Hitchcock, Friedkin was less interested in controlling narrative expectations and he preferred to affect his audience directly. According to the director, "People want to see movies because they want to be moved viscerally. . . . I mean, I'm not interested in an interesting movie. I am interested in gut level reaction. What I'm interested in is an entire audience in the palm of my hand."[3] This emphasis on a "gut level reaction" meant that Friedkin was willing to try any and every possible effect to stimulate the audience. The result was a film that worked well in this regard, but it also left a large gap in relation to the more personal films being made in the early 1970s. As film critic James Monaco astutely noted:

> As an engine of manipulation, *The Exorcist* succeeds magnificently. What other film of recent years has had the medical, psychological effect it had? It is violently effective. . . . From plot elements to special effects to

> the handling of sound (Friedkin has always been very conscious of the effect the level of the sound track has) to the nervous cutting of the music, *The Exorcist* is a catalogue of devices that work. But to what end? Technique is admirable, but eventually audiences want to hear the voice of the person who's telling the story. They may not like Bogdanovich's voice, but they can't even hear Friedkin's.[4]

Monaco's semantic choice of "an engine of manipulation" aptly described the mechanical nature of Friedkin's approach to filmmaking. Despite the lack of an authorial "voice" in a contrived and exceptionally manipulative film, there is still a highly evocative and original sound track. *The Exorcist* was one of the first films in the 1970s that sought to break down the rigid barriers between the industrial definitions of dialogue, music, and sound effects, while also actively engaging questions about the ontological nature of sound in motion pictures.

The Exorcist is unique in the evolution of film sound both for how it blurred the boundaries between sound effects and music and how it utilized extreme dynamics to stimulate the audience. Often moving from a fully modulated optical sound track to absolute silence, the film attempted to use the dynamics of the sound track to manipulate the emotions of the audience. Because the film was slated to be released in the standard Academy 35 mm optical format, all the original theatrical sound mixes were done in mono.[5] To achieve the maximum possible dynamic range from the original recordings, Dolby noise reduction was used on some of the sound effects and in the post-production mix. This was done so that Friedkin could employ radical changes in dynamic level to shock and discomfort audiences. According to sound effects recordist Ron Nagle: "Billy kept jacking up the levels till they fucked up the speakers. I tried to make everything more subtle, you know, play it down. But he kept saying, 'Jab it, stab it,' and all that shit. I'd think it was too loud, too corny, and Billy said, 'Nothing's too corny for me.' It's his style, and it makes it. It works."[6] To achieve the maximum dynamic range upon playback, Friedkin even considered releasing the film in single-track 35 mm magnetic mono to exploit the greater range of the format.[7] Although this ultimately proved impractical, as only a few theaters would be able to play a magnetic print, it marks the director's extreme concern with the use of the general loudness of the film as a tool to manipulate the audience.

During this period in the 1970s most theater owners and exhibitors adjusted the volume level in the theater to an average setting that would allow dialogue intelligibility, but without overwhelming the audience during loud passages. Friedkin's desire was to do the latter by manipulating the dynamics of the optical sound track and by exerting a nearly unprecedented control over the playback sound level. As supervising sound editor Cecelia Hall has noted,

"*The Exorcist* was one of the first films to understand the importance of affecting the audience psychologically. The director, William Friedkin, said he wanted it to be *too loud* because he wanted the audience to be slightly on edge by the middle of the film—it's very effective."[8] Although a great deal of subtlety went into the original construction of the sounds for the film, Friedkin's concern was almost exclusively focused on standardizing audience reaction: "Yes, that's how I wanted the sound, very loud. I figured it took me three months to get the sound track. It might as well be loud. As a matter of fact, I have set the sound level in each of the theaters where the picture is playing, in the 24 opening engagements."[9] By extending his purview over the film's sound track to the point of dictating the sound level within the theaters, Friedkin asserted a level of control that hadn't been seen since the days of CinemaScope and Twentieth Century–Fox's mandatory "stereo-only" policy.[10] More importantly, it signaled the start of a trend in films during the 1970s of "louder is better"—a trend that would peak with the advent of Dolby Stereo.

Friedkin also made a bold choice when he rejected Lalo Schifrin's proposed score music in favor of using several prerecorded classical and popular musical pieces in the film.[11] Although it was different from the compilation scores for *American Graffiti* (George Lucas, 1973) and *Mean Streets* (Martin Scorsese, 1973), or Stanley Kubrick's programmatic use of well-known classical pieces in *2001: A Space Odyssey* (1968), it was still a bold choice. Specifically, it went against the trend in mainstream cinema to underscore dramatic scenes whenever there was a lack of dialogue and perceptual silence. Instead, Friedkin preferred to utilize the awkward silences and pauses in the film to build the tension within the audience: "The kind of music I wanted was number one, nothing scary. No so-called frightening music. No wall-to-wall music. . . . No music behind the big scenes. No music ever behind dialogue, when people are talking."[12] Presumably the "scary" music that Friedkin referred to was that of the standard orchestral scores used for horror films, commonly utilizing tritones and other forms of dissonance.[13] While musical selections in *The Exorcist* are quite different from traditional horror scores, they are equally disturbing in their relative unfamiliarity.

According to a post-production musical cue sheet, Friedkin chose a number of established classical pieces to evoke the mood of the film. He made the musical choices by watching a silent rough cut of the picture and layering in the music.[14] The original selections included excerpts from Iannis Xenakis and Béla Bartók, but they were replaced with less familiar pieces by Anton Webern, Krzystof Penderecki, and Hans Werner Henze. What is immediately apparent in the examination of the music cue sheets is that Friedkin nearly exclusively used music only during moments of transition, when either Chris MacNeil (Ellen Burstyn) or Father Karras (Jason Miller) was coming to or leaving the house, or when Father Karras and Father Merrin (Max von Sydow) prepare for the exorcism.[15] The two main exceptions to this rule are also probably the two most recognizable musical passages from the film. The first is the use of guitarist and composer Mike Oldfield's "Tubular Bells," a semi-orchestral rock suite recorded earlier in 1973, when Chris returns from location shooting and stumbles upon Father Karras for the first time. The other piece is George Crumb's "Threnody I: Night of the Electric Insects," from his 1971 *Black Angel* composition, when Father Karras witnesses Regan's stigmata. Because each of these pieces is associated with a significant narrative event, they have both taken on thematic status in relation to the film and have been used in many subsequent horror films and parodies. This is not, however, an effect that Friedkin was likely trying to achieve. The rest of the music in the film tends toward understatement rather than serving as a leitmotif for character identification.

Much of the music in *The Exorcist* was specifically chosen and placed in the film to function on the level of liminality, being either barely perceptible or indistinguishable from general background sounds. To create these ambient musical passages, Friedkin hired Los Angeles–based composer Jack Nitzsche. Though he had previously worked as a composer for Nicolas Roeg and Donald Crammel's *Performance* in 1970 and Robert Downey's *Greaser's Palace* in 1971, Nitzsche was best known for his work in the music recording industry. He originally moved to Los Angeles when working as an arranger for producer Phil Spector and had a brief solo musical career with his 1963 instrumental hit, "The Lonely Surfer." After working as a songwriting partner with Sonny Bono, Nitzsche played on multiple sessions with The Rolling Stones in the late 1960s, and his friendship with Mick Jagger eventually led to the job composing music for *Performance*. Although he was brought onto *The Exorcist* rather late, Nitzsche contributed several musical passages that straddled the line between score and sound effect. In part this was due to his working directly with sound effect artist/composer Ron Nagle on many of the pieces. According to Cindy Ehrlich: "[Film editor Bud] Smith called Nitzsche down to compose score segments about a month before the film's deadline. Jack and Ron [Nagle] worked together creating harmonics by rubbing crystal stemware, and with dimestore musical toys. The stemware music, which Jack directed, opens the film."[16] What makes this music even more effective was how it was combined with several of the sounds created by Nagle to provide an ambient bed of sounds to accompany the images of the archaeological dig at Nineveh. Nitzsche's music was used at a number of points throughout the film as background accompaniment for scenes where spectators are kept in a state of unease because they are unsure whether they are hearing music or ambience. Although it was never articulated or theorized as such, *The Exorcist* marks one of the first moves within the Hollywood system toward the development of sound design. In this regard, the blurring of the boundaries between dialogue, music, and effects made the film extraordinarily progressive for a Hollywood production, yet the film is most often remembered for the overt deployment of its sound effects.

What makes *The Exorcist* unique in the history of 1970s film sound was the way in which any number of effects, visual and acoustic, were intrinsically tied to the supernatural aspects of the story. This was possible because the large budget for the film, as well as Friedkin's willingness to experiment with various effects, gave the sound and visual effects teams a wide latitude in the creation of new effects. Although Friedkin can be seen as initiating a series of experiments in sound effect design, sometimes his directions to the crew were expressive if not always clear. For the sound of the creatures in Chris's attic, musician Ron Nagle noted that the script called for a sound "like tiny claws scratching at the edge of the galaxy" (see figure 21).[17] When describing his conception of the demonic voice to production mixer Christopher Newman, Friedkin said he wanted the voice to sound like a Hieronymus Bosch painting.[18] To bring his desired sounds to life, Friedkin assembled a team of highly original sound effects creators. Many of the physical effects were created on the Foley stage by sound effects editors Fred Brown and Ross Taylor,[19] while Bob Fine, Gonzalo Gavira, Doc Siegel, Ken Nordine,[20] and Ron Nagle were contracted to design special sound effects for the film. Other than Fred Brown and Ross Taylor, none of the additional sound effects artists were members of the Sound Editors' Local 776 or the Sound Technicians' Local 695. Instead, each was hired as a freelance sound effect creator outside the jurisdiction of the unions. This was a bold step

FIGURE 21. A startled Chris MacNeil (Ellen Burstyn) listens to the sound of "tiny claws scratching at the edge of the galaxy" in *The Exorcist* (William Friedkin, 1973).

on Friedkin's part and one that paid off in the development of hundreds of unique sounds for the film.

Nagle, a San Francisco–based musician and record producer, was hired by film editor Bud Smith on Jack Nitzsche's recommendation to create custom sound effects.[21] Nagle had never worked in film sound before, but he crafted a number of original sounds that were used on the final sound track. Working both in San Francisco and with Nitzsche in Los Angeles, Nagle combined his musical training with an astute sense of recording technology to develop several of the now-familiar sounds from the film. In San Francisco, Nagle created numerous sounds by agitating bees trapped in a jar, getting his dogs into a fight, and recording his girlfriend's stomach while she drank water.[22] These became the basis for several of the sound effects during the film's prologue set in Iraq; however, each of the sounds was treated in the studio to estrange them from their recognizable sources. Friedkin's desire to manipulate and combine a number of the sounds in the final mix made it very important that the sound effects were recorded and transferred by Doc Siegel at Burbank Studios at "record quality" instead of "film quality" to provide the highest fidelity possible.[23] This led to the creation of a number of easily recognizable signature sounds within the film, each associated with a particular narrative event: the insect buzz of the amulet, the dog fight, the rats in Chris's attic, the bouncing bed, and, of course, Regan's head twist. The signature sounds in the film operated in a similar fashion to score music: as the sounds repeated they functioned like leitmotifs across the film, each carrying with them a certain emotional connection to a prior scene.

Because Friedkin did not want the music to carry most of the emotional weight in the film, the sound effects were used instead to stimulate the emotions of the audience. On the most basic level this was done by adjusting the film's dynamics. This is especially noticeable during the Iraq prologue where the sound track expands to its full modulation during the archaeological dig. In post-production, the sound effect stems were mixed from two interlocked three-track machines, allowing for six layers of effects. The desert scene initially includes tracks labeled "Arab chants," "digging sounds," "Moslem prayers & individual shouts," "c.u. pick axe," "b.g. sledge hits," and "break rocks" to create the foreground and background sounds of the massive dig.[24] With the discovery of the amulet, the sound track immediately contracts to a perceptual silence by eliminating the effects, music, and ambient wind, leaving just the sound of Foley footsteps. The sound that follows and engulfs the sound track, Nagle's "insect buzz" track, is one of the first signature sounds that recurs in the film and creates thematic connections between its representative scenes. Another such sound is that of the rats in the attic, a sound heard in a number of scenes during the first half of the film. Constructed from a combination of "guinea pigs running on a board covered with sandpaper, the scratching of fingernails, and the sound of a bandsaw as it flew through the air,"[25] the effect repeats several times throughout the film, each time without revealing the source of the sound. By not offering the audience a view of the creatures producing the sound and keeping their source acousmatic, Friedkin was able to make the horror of the film even more palpable by letting the audience imagine its source. Similarly, as the possession of Regan (Linda Blair) takes hold, the audience is not privy to the events transpiring behind her closed bedroom door; instead they are left to imagine the most horrific scene that corresponds to the bangs, crashes, and unearthly moans that emanate from the room.

While the sounds themselves are highly evocative and meticulously crafted, often their usage falls short for reasons that have nothing to do with the sound team. *The Exorcist* is one of the most interesting films of the 1970s to explore the basic idea of custom-designing sounds for a film, but it does so without the conceptual function behind Walter Murch's notion of "sound design."[26] Friedkin's willingness to experiment with anything and everything led to a dynamic sound track, both in terms of sounds and their volume, but he did not necessarily deploy a systematic strategy for their use across the film. Instead, *The Exorcist* is a compendium of creative sound work but without a larger system of sound usage to integrate the sounds with the narrative process. There is little subtlety in the dispersal of the sounds throughout the film and, in many ways, the film becomes anticlimactic once the exorcism begins.

The presence of the most powerful and, arguably, most "unsubtle" sound effect of the demon's voice introduces a highly conflicted element into the film: a deliberate break with the ontological nature of sound. By substituting Linda Blair's voice with any number of other voices, Friedkin harnessed the power of the sound track as a constructed artifice to affect audiences. Yet the basic premise behind the exorcism was that the audience needed to believe that this was the devil speaking through the girl's body. Although a strong effect of synchresis is created by the synchronism of the voice and lip movements,[27] ultimately the overt number and types of sounds that the demon produces often draws the audience out of the diegetic realism. The escalating use of sound effects in the film left Friedkin with no choice but to overload the demon's voice with as many acoustic tricks as possible.

According to Chris Newman, Friedkin wanted the devil's voice effect to operate as a chorus to achieve the desired demonic quality, but it was also to be heard as neutral in relation to gender, being neither male nor female but somewhere in between.[28] Initially this was done by taking Blair's original dialogue tracks and manipulating them in post-production. With Blair's voice serving as the base of the effect, Friedkin explored the possibility of utilizing other voices during the dubbing phase of the picture. According to the dubbing cue sheets from the Burbank Studios sound department, the looping process, which had recently been renamed ADR (automatic dialogue replacement), ran from 14 September through 20 November 1973, accounting for an extraordinary amount of time in post-production. Dialogue replacement for most films generally lasts between one and two weeks, depending on the nature of the film and the quality of the production recordings. Given that most of the dialogue was recorded on set by Chris Newman, perhaps one of the best sound mixers in the business, it is surprising that so much time was spent on ADR. Friedkin defended his excessive time in post-production by explaining how Blair's voice came to be replaced:

> [Linda Blair] mouthed all of those things. She had to give me a guide track. And for a long time, I thought I would use the guide track. I did a lot of experimentation with electronic distortion of her own voice and I took a lot of time and went through a lot of experiments, before I decided what to do with the voice. Briefly, most of her voice is replaced by Mercedes McCambridge, but some of her voice is her own and . . . the stuff that's the most effective was recorded in sync to the little girl's own dialogue, into a microphone as well as into a Motorola speaker. . . . The thing was literally post-synced line for line.[29]

According to Michel Chion, the powerful effect in the "mismatching" of Mercedes McCambridge's voice with Linda Blair's body became a pivotal moment for the status of the voice in cinema: "*The Exorcist* contributed significantly to showing spectators how the cinematic voice is 'stuck on' to the cinematic body. This grafting of heterogeneous elements can be seen as *The Exorcist*'s very object. Audiences could stop thinking of the voice as a 'natural' element oozing from the body on its own. . . . The whole thing takes on its horrific effect because of the relationship, the comparison the viewer makes between the visible body and the voice."[30] This heterogeneity of sound and image relations precisely marks the transition from the period of sound experimentation in the early 1970s to the reification of recording and mixing practices introduced with Dolby Stereo. Even though Chion positions the film as a progressive moment in the evolution of film sound, the ultimate result of the film's impact was to foreground multiple claims about the ontological "purity" of film sound and the proprietary value of sound effects.

Because the demon voices were constructed as sound effects in post-production rather than dialogue recorded on set, McCambridge's work was essentially effaced from the film. In technical terms, the demon vocals were married to the monophonic sound effects stems rather than the dialogue mix, making them part of the sound effects track rather than the dialogue track.[31] Thus McCambridge's vocal recordings were edited and manipulated as sound effects, physically separated from the dialogue in the editing and mixing process. Not only did this create a strange disjunction between the speaking voice and the voice that is ultimately heard in the film, but it also created a labor conflict in terms of who is acting at any given moment. Specifically, the film benefitted from the disturbing disjunction between the speaking body and the voice emanating from it, yet the filmmakers chose to credit Linda Blair's physical presence at the expense of McCambridge's contributions. This makes *The Exorcist* an extremely interesting case on the proprietary nature of sound effects, for two reasons. First, it sparked a controversy between the filmmakers and McCambridge over credit for vocalizing the demon's voice. Second, Friedkin claimed that the film's sound effects were the legal possession of the studio, which led to a well-documented court case against the makers of the Italian horror film *Beyond the Door* (*Chi sei?*, Ovidio G. Assonitis [1974]; U.S. release 31 July 1975) over the question of the ownership of sound.

McCambridge was not credited for her contribution to the film upon its initial release in 1973, and the suppression of her screen credit was done in part to prevent harming Blair's chances of receiving an Academy Award nomination for Best Supporting Actress. It was only after the Oscar ballots were tabulated in late January 1974 and Blair received the nomination that news leaked to *Variety* and *Time* about Mercedes McCambridge's contribution.[32] In an interview with Charles Higham of the *New York Times*, McCambridge recounted her contribution to the film: "Doing that sound track was a terrible experience. I just didn't do the voice, I did all of the demon sounds. That wheezing, for instance. My chronic bronchitis helped with that. I did it on one microphone, then on another, elevating it a bit, then a third and forth, two tones higher each time, and they combined them, as a chorus."[33] According to the log sheets, after weeks of experimenting with Blair's voice McCambridge was called in to provide "Regan's Devil Voice" in six sessions lasting from 10 to 18 October.[34] Upon the completion of McCambridge's dubbing work, Friedkin and his team embarked on dialogue and effects mixing at the Todd-AO and Twentieth Century–Fox sound departments from 22 October through 15 December. While ADR sessions were occurring, Friedkin supervised the mixing of McCambridge's vocal contributions, and his notes include instructions to slow down certain sounds, switch between the "demon" and Regan's voice, build tracks from multiple takes, overlay multiple voices, and create the "backward" voice by reversing the tape. Despite the manipulation of McCambridge's voice as an effect, the particular raspy qualities of her voice are quite recognizable and her vocal performance lends a great deal of weight to the creation of the demon.

In an interview with the American Film Institute in early 1974, Friedkin admitted that he chose McCambridge precisely for the desired "emphysemiac" wheeziness of her voice, a sound that was used prominently whenever the demon was not speaking.[35] Simply by listening to the film with the knowledge that McCambridge provided elements of the demon's voice, it becomes apparent that she made a huge contribution to the overall film. But perhaps one of the most disturbing aspects of the film's production process is in Friedkin's response letter to Charles Higham, where the director defended his decision not to give McCambridge screen credit: "This is a common practice in the film industry. The best compliment an actress can receive for dubbing a voice is that audiences are unaware of the dubbing, and for obvious reasons, studios don't go around publicizing what was dubbed or by whom any more than a magician makes it a practice to publish the tricks of his trade."[36] What Friedkin did not acknowledge is that he chose McCambridge's voice precisely because of its recognizability as well as its ambiguous gender and guttural qualities. Hidden in his statement is an unusual bias that haunts *The Exorcist*: the specter of the industrial split between dialogue and sound effects recording. Friedkin's reaction to McCambridge's contribution exposed several of the abiding assumptions about film sound that still operated in the 1970s. First, that the voice of the speaking body was intrinsically the "real" voice, while the voice being added in dubbing was somehow an extra effect. Despite the fact that a large portion of the dialogue in the film was replaced in post-production, Friedkin established a precedent where the voice of an actor, whether recorded live on set or in ADR, was somehow the only voice authorized to match to his or her body. Second, because the effects in the film were created instead of generated by the objects to which they are attached, it was assumed that they held a proprietary value.

This was evidenced in September 1975 when Friedkin and Warner Bros. brought suit against the studio responsible for *Beyond the Door*, alleging that the film copied and imitated several of the signature sounds created for *The Exorcist*. In the suit, the litigants filed claims that they were entitled to reparations because "the creation, development and execution of the sound effects in *The Exorcist* was a monumental task extending over many, many months at a cost of several hundred thousand dollars."[37] Claiming that the sound effects were copied by the Italian film did not mean that the actual effects were electronically duplicated, but that they were emulated and that the sounds were "arrived at only after the Italian film makers had 'studied and dissected' the effects achieved in *The Exorcist*."[38] What Friedkin and Warner Bros. sought to demonstrate was that the originality of *The Exorcist*'s sound effects in fact made them proprietary and therefore protected under copyright law. Yet instead of attempting to sue the Italian filmmakers for copyright infringement, the lawsuit was filed to prevent the distribution and exhibition of *Beyond the Door*, which was seen as both competition to the value of the film franchise and "merely a cheap imitation of *The Exorcist*."[39]

But the most disturbing aspect of the lawsuit is the way that it completely dismissed the special contributions of the sound effects artists and musicians. Oddly, the suit cited only three sounds that were emulated in the Italian film: the demonic sound of the loud scratching, the voice of the devil within Regan, and the multifaceted voice of the devil.[40] Not only was the first sound effect the legally contested and uncredited contribution of Ken Nordine, but neither Friedkin's nor Smith's affidavit gave any mention of McCambridge's significant contribution. This lack of credit echoes the way Friedkin and Warner Bros. did not recognize the contributions of all those involved in creating the sound effects for *The Exorcist*. Ironically, the lawsuit excoriates *Beyond the Door* for bearing "the heavy hand of the copyist" in re-creating the effects,[41] which implies that the sounds created for *The Exorcist* were both original and essential in advancing the story. Although the use of non-film industry personnel to create many of the sound effects was unique, the fact that the effects designers were hired as contractors and had no input regarding the use of the sounds indicated that little had changed in the separation of labor in Hollywood. Hence the "heavy hand of the copyist" was possible only because of the superior quality of *The Exorcist*'s sound effect work, yet because of the way the contributors were paid as freelance employees and not given screen credit, much of the labor behind the sound construction was hidden from the public.[42]

Ultimately the advances in sound technique and sound effects creation in *The Exorcist* were overshadowed by Friedkin's single-minded desire to stimulate his audience by whatever means available to him as a director. Even though the creation of the numerous sound effects and musical elements of the film was truly produced on an unprecedented scale, the end result of their use was more to manipulate the audience rather than to augment the story. This carried over to the film's use of obscenities, and as William Van Wert commented, "If there were no special effects in the film, no blood and gore and visual violence, the film would still shock audiences, because the language Chris uses would shock."[43] Nearly every device in the film, from its makeup and prosthetics to its language and sound effects, was calculated to have a maximum impact on the audience. Unfortunately, the result was that the careful work that went into the creation of the sounds for the film was drowned out amidst the cumulative distraction of the other effects. What once had the potential for being a taut psychological and supernatural thriller became a compendium of effects used to manipulate audiences and generate box office success. Or, to invert Friedkin's comment that "there wouldn't be a *Mean Streets* without an *Exorcist*," it is safe to say that the success of *The Exorcist* made it increasingly difficult to create personal films like *Mean Streets*. The years that followed *The Exorcist* witnessed studios embracing the "blockbuster" approach to filmmaking and an ever-growing gap between box office successes and personal films. The process of custom sound effects design found a home in the science fiction and horror films in the mid- and late 1970s and in the rise of Dolby Stereo as an exhibition technology.

Moving away from an early emphasis on music-based films, Dolby Laboratories found their first major successes in the utilization of Dolby Stereo for a number of late 1970s science fiction films:[44] *Star Wars* (George Lucas, 1977), *Close Encounters of the Third Kind* (Steven Spielberg, 1977), *Invasion of the Body Snatchers* (Philip Kaufman, 1979), and *Altered States* (Ken Russell, 1980). The transient effects that proved to be bothersome in the earliest version of 35mm Dolby SVA were partially tamed by 1978 with the change from the Sansui QS matrix to the MP matrix, which provided better channel separation.[45] Yet even with the occasional sonic artifacts, the flexible codes of realism in the science fiction genre accommodated the limitations of the Dolby system. In the hyperbolic diegetic worlds of sci-fi, acoustical ephemera could be assimilated into the form without disrupting the nature of the narrative. Arguably, Dolby SVA's lack of divergence and shifting sound field could be an asset to the spectacular nature of the science fiction genre, specifically, the use of the surround channel to introduce narrative elements in the space of the theater and the ability to surround an audience with ambient sound. Similar to its ability to reproduce stereophonic music in its full dynamic and frequency range, Dolby Stereo formats began to be associated with its spatial effects in relation to science fiction films. But just as the genre assimilated the use of Dolby Stereo, so too did it rapidly impress its own representational codes onto the technology. After the inspired use of the surround channels to convey the writhing horror of the pods from the *Invasion of the Body Snatchers* or the alien drag races in *Close Encounters*, future uses of the surround channels were regularly entangled with either the portent of the sinister or the uncanny.

Star Wars (George Lucas, 1977)

If a single film is most closely associated with the promotion of Dolby Stereo technologies, and, by extension, with the development of a Dolby Stereo aesthetic, it is George Lucas's *Star Wars*. It should be noted, however, that the techniques and practices used on the film followed several standard film sound aesthetic choices and attempted to make them fit the Dolby Stereo technologies. The result was a motley mix of sound effect innovations, familiar cinematic devices, and astute marketing strategies that made the film the most successful box office draw of the 1970s.[46] Yet amid the disparities among form, content, and technique there was a serendipity that has positioned *Star Wars* as an epochal achievement in the history of American film. One way that the film impacted cinema history was by convincing studios that the youth market was still a viable engine for cinematic success. Another result was the extreme efficacy of advertising and cross-promotional marketing to generate a large audience and to sustain that level of publicity throughout the film's entire run. A major tool in maintaining the promotion of the film was the connection between the film's use of Dolby Stereo and the audiophile culture of American youth.

Lucas learned with *American Graffiti* (1973) that the careful marketing of nostalgia could reap great rewards by tapping into a submerged cultural memory. In the same way, he positioned *Star Wars* as a "space opera" that would reference not only the Flash Gordon serials of his youth, but also a whole range of cultural allusions from *The Searchers* (John Ford, 1956) to *Casablanca* (Michael Curtiz, 1942) to *The Hidden Fortress* (Akira Kurosawa, 1958). The cultural referents of Lucas's youth would not necessarily translate for the teenagers of 1977, however, and the decision to release the film in Dolby Stereo marked an effort to interpellate an audience of audiophiles familiar with the function of Dolby noise reduction. Dolby Laboratories sought to promote the effect of their technology to consumers rather than its operational characteristics. Instead of advertising how the technology reduced noise, Dolby promoted their brand as a form of sound enhancement. The strategy was effective enough to make Dolby an instantly recognizable name by 1977, and though audiences had little or no idea how Dolby Stereo worked, they were convinced that it would make movies "sound better."[47] The paucity of marketing information actually worked to Dolby's advantage because it set up few preconceptions of how the technology was supposed to be deployed in favor of a *tabula rasa* for the creative filmmaker. Nevertheless, in the case of Lucas and *Star Wars*, certain preconceptions were mobilized in the utilization of the technology, and other established practices were imposed on the technology due to its incommensurability. The result was a lopsided film sound aesthetic, one that simultaneously presented new ideas of film sound use while reifying several established mixing practices and representational strategies.

In his seminal essay on the development of Dolby Stereo, "Une esthétique Dolby Stéréo," Michel Chion pointed out a number of the aesthetic properties involved with the technology. Chion demonstrated that the technology was used regularly to create three specific effects: a hyper-realism of richer, sharper, and more precise sounds; a reduction of all sounds to the "present indicative" of progressive sound; and a shift in the status of offscreen sound from "elsewhere" to "nearby."[48] Chion was most interested in how Dolby Stereo functioned to change the ontological status of sound effects by increasing their fidelity and thereby providing a form of "sensory authentication" for the visualized source. The result was that often the overtly present quality of sound effects in Dolby Stereo forced sounds to the foreground, making it much harder for them to operate suggestively or symbolically.[49] Chion noted that in Dolby Stereo, "One is in the prolific kingdom of *immediateness*, the here and now. Rare are the films in Dolby Stereo that articulate a symbolic temporality through sound."[50] Part of what was lost in the Dolby Stereo aesthetic was a sense of ambient temporality, a continuous flow of time amidst the plurality of transient sound effects.[51] This manifested itself in *Star Wars* through two very central practices: the use of nearly continuous score music and the regular dubbing and replacement of production dialogue. The first practice was related in part to the second as a way to cover the resulting gaps in background room tone and ambience. These two practices were linked to several divergent assumptions about Dolby Stereo and how they dictated the evolution of film sound in the late 1970s.

The main assumption about Dolby Stereo sound was that it was a technology designed first and foremost to make film music sound better. Of course, this was a common misperception based in the rhetorical strategies of Dolby Laboratories linking their film sound system to the ubiquity of their noise reduction technology. Although noise reduction was a major component of the Dolby Stereo system, it was only a part of the overall function of the technology. What was more significant was the way in which Dolby Stereo was able to spread the sound across the screen in a stereophonic field and also extend the sound into the space of the audience through surround speakers. The result of using 35 mm Dolby SVA for film music was that it replicated the effect of hearing the music as though performed in the space of the theater. A drawback, however, was that audiences were conditioned to hear the overt "presence" of the music but had to listen *through* the music to hear the dialogue and sound effects in the mix. This is evident from the very beginning of *Star Wars*, where John Williams's prologue music first accompanies the scrolling text introduction and then ducks down in volume to allow the main action to unfold. Because of the music's ability to fill the auditorium with sound, Lucas had to top the opening music with an even more spectacular sound effect: the roar of the starships as their sounds pass from the surround speakers to the front speakers.[52] Impressive as it may have been, Lucas was never entirely willing to let go of the score music that accompanies the introductory scene, a technique that he repeats throughout the film, creating a wall-to-wall use of music.

This tension between the deployment of sound effects in the aid of narrative exposition and the use of score music to provide emotional cues for the audience was perhaps the most disturbing aspect of the sound aesthetic being forged in *Star Wars*. Lucas was trying to be a technological and stylistic innovator with his sound effects while simultaneously wanting the emotional pull of a classical Hollywood musical score. This resulted in a clash of representational strategies throughout the film where sound effects, music, and dialogue compete for a central position in the soundscape. Perhaps the best way to examine this phenomenon is to consider the "highs and lows" of the film. This not only refers to the narrative crescendos and denouements, but also the expanded dynamic range of the sound track and the resultant conflict between sounds. Quite often Lucas would underscore a particularly dramatic moment with music while also including a wide array of sound effects in the mix. Though the stereo spatialization of the sound effects aided in increasing the clarity of the mix, the score and sound effects often were placed in the same musical range. An example of this occurs during the opening sequence when the Imperial cruiser is boarding the rebel ship. It is clear that the effect of the smaller ship being "swallowed" by the larger one was meant to be conveyed through sound from the first line of dialogue—"Do you hear that?"—and the way that the characters stare at the ceiling before the film cuts to an exterior shot. However, at the same time that the sound effects are working to convey the enormity of the external event, the score music is also playing, partially masking the sound effects. This redundancy carried over to the next narrative event when the storm troopers cut their way through the blast door. Once again the allusive sound effects, which fully convey the effort being exerted, are undercut by the ubiquitous presence of Williams's score. More disconcerting is that the higher frequency sound of the strings and the sound effect of dismantling the door are in the same acoustic register and in conflict with one another. Lucas left himself nowhere to go dynamically or tonally than to back away from the plurality of sounds, only to build them up again and again.

This method of filling the sound track's dynamic range was deployed throughout the film and led to regular conflicts with the development of the narrative. Because Lucas was unwilling to let go of the security of the music track and allow the sounds of the film do the work for him, he often layered too many sounds on the sound track. In a standard monophonic mix the conflict between the two tracks would have been immediately apparent in a loss of clarity, and measures would have been taken to filter a particular frequency band in the music to make a space for sound effects and dialogue. Yet with the expanded dynamic and frequency ranges of Dolby Stereo, not to mention its spatial characteristics, the pragmatic techniques of monophonic film sound mixing were largely forgotten.

Another common technique that was employed in post-production sound was the coordination of sound effects work with the scoring of the film to ensure that one did not interfere with the other. Yet because John Williams spent over a year preparing the score for *Star Wars* before recording it in March 1977, the same time period when the sound effects were being edited, there was an inevitable conflict between the two tracks in the final mix. In order to achieve any dynamic changes it became necessary to build a scene to

FIGURE 22. Alerted by offscreen sound effects, Princess Leia (Carrie Fisher), Han Solo (Harrison Ford), and Luke Skywalker (Mark Hamill) realize they are trapped in a giant trash compactor in *Star Wars* (George Lucas, 1977).

its maximum volume, cut away, and then return at a lower volume to start increasing it again. This was the mode of operation during many of the battle sequences, but its shortcoming carried over to scenes constructed without music. In their escape from the prison, the protagonists become stuck in a garbage chute only to find out that the room is in fact a giant trash compactor (see figure 22). This is made apparent when the sound effect of the room's hydraulic system kicks in to convey the direness of the situation. After the machinery switches on, the rhythmic pulse of the sound effect increases in tempo and frequency, creating urgency in the auditor. However, because Lucas increases the tempo of the effect so quickly, he is forced to cut away and return with the tempo noticeably slowed. In total, the sequence repeats this motif three times, thereby undercutting the initial urgency of the scene and drawing attention to the clumsy use of the sound effect.

That is not to say that sound effects do not play a large part in the creation of the other worlds and their inhabitants; as Gianluca Sergi noted, sound regularly was used to "audilize" several aspects of the film's narrative and to create an "auditory memory" of a world that the spectator has never encountered.[53] Once Lucas decided to ignore the basic physical laws that dictate that there is no sound in outer space, he was free to use sound effects as liberally as he wished.[54] In hiring Ben Burtt, fresh from USC film school, to create sound effects for the film, Lucas took an extraordinary step in his commitment to using sound as a major storytelling partner in the film. Burtt's monumental contribution to *Star Wars* came not only in the conceptualization and design of sound effects for the film, but also through his attention to the careful use of each sound. Nuances in sound variation allowed alien characters to "talk" and be recognized by the audience, not because of the semantic elements in their new languages, but through the expressive qualities of their voices. Burtt was also known for the creation of hundreds of signature sound effects for nearly all the sourced sounds heard in the movie. However, the creation of these effects was entirely dependent on Lucas's script, which indicated what sounds would be needed and in what narrative context. Whether the clicking of an aqualung respirator for Darth Vader's breathing, the servo-motors of the androids, or the electric crackle of the light sabers, each sound effect was introduced in relation to its moment of narrative significance. Although this made for a remarkable compendium of effects, they were tethered to the central narrative development with little autonomy of their own. Very rarely is an object or character introduced by sound first; instead, nearly every sound source is visualized at the same time the acoustic effect is heard. And even when it was possible to use a sound advance to create dramatic tension, as in the film's first sequence, Lucas highlighted the sound effect through descriptive dialogue that verbally pointed to what the audience already heard.

The main question behind these observations is simply why Lucas chose to build such redundancy between the sounds and images into the film. Or to reverse Sergi's prior equation, why did Lucas choose to visualize nearly every one of the sounds that was created for the film? In part it was indicative of an ongoing mistrust of sound as a storytelling medium. This is evident in how the film tries to hide the heavy use of dialogue looping and the very uneven dialogue backgrounds behind the nearly ubiquitous score. The refined dynamic range and precision of Dolby Stereo made previously inaudible changes in dialogue background levels audible and therefore the disparity in room tone is far more noticeable when looped lines are intercut with production sound or when there are variations in location sound backgrounds. Lucas and his sound team dealt with this by regularly keeping the score music at a low level in most scenes, but when the score was absent, the audience became aware of the additional absence of ambience. One particularly shocking example is Luke's (Mark Hamill) conversation with his aunt and uncle over the breakfast table. Because the scene was filmed on an outdoor location and the background ambience changes according to the shot and the microphone position, two elements of the sound become immediately apparent. First, when Luke and his uncle (Phil Brown) are conversing, each of their voices carries the reverberant spatial qualities of the room. This spatial signature is missing from nearly every other scene because most of the dialogue was replaced in ADR with no attempt to restore the missing room tone. This is overtly noticeable when Luke's aunt (Shelagh Fraser) speaks. Her voice was dubbed in post-production and carries no spatial characteristics, so as the film cuts from one character to another the backgrounds change radically, even to the point of a complete absence behind the aunt's voice. Second, the angle of the miking also dictated the amount of background noise that was picked up. Because Luke was filmed against a wall, there is more reverberation but fewer external noises on his track. But because his uncle was filmed with an open doorway behind him there is less reverberation in his voice and a greater proportion of background sounds, including bird chirps, heard on the sound track.

The incommensurability of the backgrounds behind these dialogue recordings points to several curious strategies employed by Lucas and his sound team. The main shortcoming in the film's sound was that despite the careful attention given to hard sound effects, such as weapon noises or alien languages, there was little thought given to an overall sound aesthetic. In part this was due to the lack of a central organizational figure overseeing the entire sound construction from production through post-production. Unlike Walter Murch's function on *The Conversation* (Francis Ford Coppola, 1974) or Jim Webb's connection to the post-production process on *Nashville* (Robert Altman, 1975), Ben Burtt's work on *Star Wars* ended when the sound effects were edited to the visuals. Moreover, re-recording mixing team members Bob Minkler, Don MacDougal, and Ray West were relatively new to post-production mixing and all were working in Dolby Stereo for the first time. This resulted in an attempt to "fill up" the stereo field with as many effects and sounds as possible to keep the film from sounding thin in its multichannel playback. This was generally done by using music whenever possible to hide the spatial lack involved in using monophonic dialogue and effect source materials. While this did work well for the discrete channel 70mm mix, where monophonic sounds could be reinforced by adding divergence, it caused a major problem with the 35mm optical release in Dolby Stereo.

Specifically, the problem with the lack of divergence was introduced when mixing the film for a 35mm Dolby SVA release. Divergence allowed the mixer either to position a monophonic sound as a "point source," where it had a specific location within the stereoscape, or to "spread" the sound across several speakers, often to expand it into a background ambience.[55] Re-recording mixer Richard Portman was one of the first in the United States to work in Dolby SVA, and he noticed a radical difference between mixing sound effects for 70mm discrete magnetic sound and in 35mm Dolby Stereo optical:

> What [divergence] did was simple. . . . You could take one track and you could spread it through all the speakers. Because it was recorded in its own discrete channel, there could be no phasing within the channels, and you could get it out in the theater and you could fill the room with one track. . . . Not so in the Dolby world; you have to have dissimilar sounds in the channels. If you want to have a nice stereo feel it's fine. But if you take a monophonic track you have to double punch it, left and

> right, and then what happens is it seems to be a center image, you don't get any spread on it. Instead of a monophonic background track, now you have to provide a stereo track to establish the basic background.[56]

In the case of *Star Wars*, this meant that in mixing the original four-channel discrete mix through the Dolby Stereo matrix, any monophonic sound effects that were expanded across the stereo field would collapse back to mono upon decoding. This became a major problem and one that was generally solved by further utilizing the stereophonic score music to fill the auditorium and cover the collapse of sound effect elements in the soundscape. In this regard *Star Wars* is an excellent example of several conflicting characteristics of the "Dolby Stereo aesthetic." First, the precision of sound reproduction and extreme dynamic and frequency ranges often revealed the limitations of the production recordings and the mismatch of sound over cuts. Second, the matrix at the heart of the Dolby SVA system had a tendency to re-channel several sounds to the center of the screen instead of their original location in the discrete four-channel mix. And third, the problems of the Dolby Stereo system were often hidden behind the overt use of score music as a method to distract the spectators from the system's basic shortcomings.

Finally, there was a major distinction about Dolby Stereo that was not expressed to the public in either the advertising or the trade presses: the difference between Dolby Stereo for its 70 mm and 35 mm releases. Because the term "Dolby Stereo" was used in advertisements to refer to both the 70 mm and 35 mm formats, audiences were regularly unsure what format they were going to hear when watching a film. Moreover, the fact that several unscrupulous exhibitors would use the Dolby logo to advertise that the print was recorded in Dolby Stereo, even if it was being played back in mono, led to further semantic confusion over the system. In order to capitalize on the wave of publicity behind the Dolby Stereo technology, Lucas and his producer Gary Kurtz convinced Dolby Laboratories to refine their six-channel 70 mm system to include an exclusive "baby boom" bass extension channel. Unlike Sensurround, this was not a sound generator that produced low frequency rumbles in response to a guide track, but rather a method for augmenting low frequency reproduction from two dedicated discrete channels of audio information. Because the stereophonic spread of the five screen channels of 70 mm sound could easily be reproduced through only three channels, Dolby Laboratories decided to eliminate the left-center and right-center channels in the 70 mm format and replace them with additional low-frequency information below 200 Hz. This allowed the system to expand the bass response for theaters wired for six-channel 70 mm sound by using the existing speakers to play back the low frequency information without any additional rewiring or new hardware.[57] The low-frequency enhancement was not only a response to the recent success of the Sensurround system to reproduce ultra-low frequency information, but it was also done to provide a quantifiable acoustic effect. Although the Dolby documentation promoted a "more life-like sound, especially on music and effects,"[58] the low-frequency enhancement was also added to draw in a new audience of audiophiles to experience the effect as a physical sensation.

The new 70mm Dolby Stereo format was an attempt to tap into the youth market through the most ubiquitous audiophilic device: the home stereo. Dolby Laboratories was less interested in providing an expanded dynamic range that closely emulated the natural range of hearing, as they had with their professional noise reduction equipment, than with targeting a specific demographic with their sound technology. Since the demands of screen realism were loosened in the science fiction genre, it was less important to have an accurate reproduction of previously unheard futuristic sounds than it was to foreground their spectacular nature. In showcasing the originality of the sound effects, not only was the expanded volume range of the system important, but so was the ability to emulate the low-frequency bass effect of many home stereo systems. By doing so, the film catered to a particular audience interested in deriving the same low-frequency effects from their film choices as they did from their musical choices. This inclusion of the low-frequency channel, in addition to the surround channel, further expanded the effect of immersing the audience in sound. According to Olen J. Earnest, director of marketing research and analysis for Twentieth Century–Fox, "The majority (of respondents) described the sound as taking the individual moviegoer beyond the passive moviegoing experience, and into a more active participation."[59] This was further echoed in the sentiments of National Association of Theater Owners technical advisory committee chairman Al Boudouris, who noted that "the sophisticated electronic techniques for reproducing strange, new good sound and space oriented noises is an extra plus that can greatly enhance all pictures and make sound in motion picture theatres sound better than from a television set or home hi-fi system."[60] More than the film industry trying to differentiate itself from television or home audio, something it had done since the 1950s, the creation of the "baby boom" channel in 70mm Dolby Stereo was an attempt to poach one of the main quantifiable aspects of home stereos and add it to Dolby cinema sound as low-frequency enhancement.

While this did target and capture a particular audience through its demographics, there is a larger question at stake in the examination of *Star Wars'* reception: exactly what did people really hear when they went to see the film? Although this may seem a highly subjective question, it is meant to address the problematic issue of conflating Dolby Stereo's 70mm six-channel discrete magnetic sound with 35mm optical Dolby SVA. Dolby Laboratories were heavily invested in the success of *Star Wars* to promote the efficacy of their 35mm optical Dolby SVA system. According to Dolby representative George Finkhausen, "[The] biggest problem selling exhibs on a Dolby system is convincing them that the Dolby set-up is not just another sound gimmick."[61] However, the way that Dolby Laboratories promoted the SVA system was on the back of the success of the 70mm format. Although the SVA system did provide an improved sound quality over the monophonic optical format, it had neither the same spatial characteristics nor the dynamic range of the 70mm system. As Finkhausen pointed out in *Variety*, "In general, demographics of the frequent film patron matches up with the demographics of the group that supports the bulk of the music and hi-fi industries. This public wants good sound quality when it goes to films."[62] Even though the 70mm Dolby Stereo format was designed to deliver the best possible sound reproduction to audiences, it was not necessarily what most people heard when they saw the film in theaters.

Out of thirty-two theaters showing *Star Wars* in its opening week, eighteen were ready to show the film in Dolby Stereo.[63] In total, only forty theaters in the United States were equipped with Dolby Stereo equipment by June 1977.[64] This meant that despite the fact that *Star Wars* was the first film to go into wide distribution in Dolby Stereo, both in the 35mm and 70mm formats, most theaters were playing the film in standard monophonic optical. In fact, the plurality of formats led to a great deal of semantic confusion in regard to what audiences were actually listening to. According to a special issue of the *International Newsletter about 70 mm Film* on Dolby Stereo, "Except for a few rare 70mm show-prints . . . used in major first run cinemas in Los Angeles, New York and London, the [35mm] sound was now 'folded in' in such a way that most dialogue came from the center speaker."[65] Although there is no empirical evidence that cites the success of the 70mm format in relation to the 35mm format, it is clear that only a few audiences were reacting to the expansive sound of the 70mm format in the first weeks of the film's release. As the film trickled into smaller theaters, most without Dolby Stereo, the word of mouth had convinced audiences that they were seeing, and hearing, a quantum leap in cinematic representation. This was primarily due to the spin that the trade press, Twentieth Century–Fox, and Dolby Laboratories used in marketing Dolby Stereo as a panacea for the film industry. *Variety* was able to gain access to a "confidential" internal survey from Twentieth Century–Fox that demonstrated the success of Dolby Stereo and *Star Wars*. According to an article from 17 May 1978: "*Variety* has learned that Fox discovered, to its apparent surprise, that 'those who saw "Star Wars" in stereo tended to rate the picture higher in all its elements.' Of those responding to a random telephone survey last July in cities where 'Star Wars' played in Dolby Stereo, '90% said it added depth to the picture, which was rated equally as important as the visuals.'"[66] Once again, the semantic imprecision of the survey, citing that the sound added "depth" to the picture but not whether that meant acoustic depth or story depth, points to a general misunderstanding of the purpose of the technology and its correlative aesthetic effects. However, what it did demonstrate was that audiences found more value in the picture when they were aware of the fact that they were listening to it in Dolby Stereo, regardless of what that format may have been.

This led to a boom in cinema sound enhancement in the wake of *Star Wars*' success. Driven by Dolby Laboratories, but challenged by ephemeral theatrical sound competitors like Eprad and Kintek,[67] multichannel cinema sound took hold in the late 1970s on a level unheard since the 1950s. With the release of *Close Encounters of the Third Kind* as the first film in Dolby Stereo 70 mm magnetic and Dolby SVA only, Dolby Laboratories were able to prove the backward compatibility of the SVA format to doubting exhibitors. As with *Star Wars*, the 70 mm release of *Close Encounters* in November 1977 was used to create a market demand for the SVA version. Once again, the dissemination of the Dolby SVA format was entirely dependent on the promotion of 70 mm Dolby Stereo sound aesthetics. Hence, when most audiences went to the theater to hear their favorite films in Dolby Stereo they were generally hearing it in the 35 mm optical format.

In a telling neologism, science fiction author and perspicacious film reviewer Ursula K. Le Guin christened the effect of Dolby's new boom channels as "decibellicocity."[68] In her review of *Close Encounters*, the author correctly noted that the sound track was often used to "whip up emotions, the same trick that's so easy to do with electronically amplified instruments."[69] This "boom aesthetic" had far more to do with emulating the wide-band loudness of home stereo systems and rock concerts than it did with accurate reproduction. Whether the opening dust storm in *Close Encounters* or the first appearance of an Imperial Destroyer in *Star Wars*, the main impression left with the audience was not the subtlety of the sound editing but the all-encompassing surround sensation and the extreme volume of the event. No reviewers wrote about the full-frequency reproduction of John Williams's orchestral score or the lack of distortion and ground noise; instead, the emphasis was on how Dolby Stereo, particularly the 70 mm format, allowed filmmakers to "pack in" more sound.

Thus the idea of a Dolby Stereo aesthetic, as put forth by Michel Chion and several other authors, is one that is fraught with contradiction because of the synchronic duplicity of the technology and its diachronic change throughout the 1970s. In the case of *Star Wars*, the aesthetic properties that emerged from a combination of increased background noise on production tracks, imprecise editing techniques, residual mixing practices, and technological limitations became the template for many of the future films made in any Dolby Stereo format. The gap between the actual sound of Dolby Stereo, whether in the 70 mm or 35 mm formats, and the idealized perception of its potential would take several years to close. It wasn't until the highly theorized and rigorously structured sound tracks from directors such as Wim Wenders, Jean-Luc Godard, Jonathan Demme, and the Coen brothers that Dolby Stereo came into its own in the late 1980s and early 1990s as a powerful device to aid in the storytelling process. That is, with three critical exceptions: the unparalleled sound work of *Days of Heaven* (Terrence Malick, 1978), *Apocalypse Now* (Francis Ford Coppola, 1979), and *Raging Bull* (Martin Scorsese, 1980).

10

The Sound of Storytelling

Dolby Stereo and the Art of Sound Design

Despite the fact that a number of preexisting sound roles gained official recognition during the 1970s, including those of supervising sound editor and sound effects editor, the labor system was still an established hierarchy of differentiated responsibilities, and the division of labor between production and post-production remained nearly complete. In contrast to the fixed and regulated rules of sound effect construction and mixing that came in with Dolby Stereo, a new group of sound sensitive practitioners and directors sought to redeploy and retool the technology for creative uses. This led to a series of aesthetic experiments in the 1970s, culminating with the emergence of the sound designer as the individual tasked with conceptualizing the overall sound of the picture.

Dolby Stereo, Part Two: Storytelling

One of the biggest problems with examining Dolby Stereo technology is that one is actually examining a *nested series* of technologies. This is an error that was aggravated with each new generation of Dolby Stereo in the 1970s and the 1980s. Perhaps the individual who should be praised for first theorizing Dolby Stereo is French film critic and theorist Michel Chion. Writing in *Cahiers du Cinéma* in the early 1980s, Chion put forth several loose doctrines on Dolby Stereo technologies and offered poignant insights into their aesthetic natures. Unfortunately, Chion's attention to the presentational aesthetics associated with Dolby Stereo resulted in a moment of historigraphic blindness. Most of the films referenced in Chion's articles were released in discrete-channel 70 mm format rather than the matrixed stereo used on the 35 mm optical prints. Because of this, Chion failed to identify the separate operational methods of the different technologies and post-production mixing practices involved. His missing analysis of the technologies aside, he did provide a number of interesting assertions that can be useful in considering theories of multichannel sound in general and Dolby Stereo in particular.

In "Une esthétique Dolby Stéréo," the three properties of Dolby Stereo that Chion identifies apply to both Dolby SVA and Dolby 70 mm. As previously mentioned, the second function of Dolby Stereo is the way that sound may be said to be more present and utilized "progressively."[1] Perhaps a better way of phrasing this is that the sound track tends to be filled with ever-changing sounds, rarely pausing or becoming static. It is here that Chion identifies a latent potential in both Dolby systems, the ability to construct cinematic "environments" through sound rather than just mise-en-scène. Early attempts at this can be heard in the shipwreck and deserted island in *The Black Stallion* (Carroll Ballard, 1979) or the sonic miasma of *The Elephant Man* (David Lynch, 1980), though perhaps its finest manifestation first occurred in Terrence Malick's *Days of Heaven* (1978).

If previous reviews of Dolby Stereo noted its capacity for spectacle, *Days of Heaven* was the first film where critics took note of Dolby's capacity to enhance storytelling. In particular, the hyperrealism of Dolby Stereo's reproductions let Malick and his sound crew foreground the pellucid sounds of nature to root the story in the Texas Panhandle in 1916. By way of contrast, industrial Chicago, with its harsh metallic sounds and deafening steel mills, is set in counterpoint to the expansive soundscape of a Texas ranch during harvest season. As Charles Champlin described it upon its release, "The most spectacular sequence, a wheat harvest, a plague of locusts, a prairie fire at night (an awesome piece of film-making) will stay in mind forever, and so will its sounds and its silences. Its summoning up of the resources of sight and sound is without parallel in this or many recent years" (see figure 23).[2] Or, as Chion observed in retrospect, "[Dolby] multitrack sound allows for vastly more present audio impressions of nature, largely by opening up the audio field to the high treble: the rustle of insects, the buzzing of flies, the high tweets of certain birds, the audio background of landscape acquires more presence, even if it is still far from perfect fidelity to nature."[3] Indeed, the detail of sound effects in the film is unique and the sound track itself takes up much of the cinematic storytelling, with voiceover often in lieu of direct dialogue and a complex interplay of sound effects, multichannel ambiences, and music, both diegetic and nondiegetic. Chion singled out the use of voice in the film in particular, noting how "human speech is constantly being reinscribed into the general clamor of the world with its noises and vibrations, often by a tight braiding of moments with speech and moments without, sensations of noises and words."[4] This synesthetic effect was even showcased in the advertisements for the film, which promised, "Your eyes . . . Your ears . . . Your senses . . . will be overwhelmed."

FIGURE 23. Terrence Malick used Dolby Stereo technology in *Days of Heaven* (1978) to surround the audience with the ambiences of steel mills, wheat harvests, and a plague of locusts.

Yet Chion's most powerful observation in his evaluation of Dolby Stereo is the changed status of stereo in the Dolby system. Where the discrete channel separation of the earlier stereophonic systems allowed for a sound to inhabit one of the lateral speakers, making possible the use of active offscreen elements, Dolby SVA's limited separation can only provide a "passive offscreen" presence. According to Chion, the sounds are "no longer 'elsewhere,' but 'nearby,'" hovering in the wings threatening to reveal themselves.[5] He further states that "the sound has done poorly in finding its symbolic anchorage, its structural function in the mise-en-scène," because the "hyper-realistic aesthetic" of the sound makes plain the flatness of the visual presentation.[6] Simultaneously hyperreal in its ability to render the sounds with exacting clarity and surreal in the compression of the lateral width of the screen space, Dolby Stereo presents an aesthetic that draws attention to the lack of the image, Chion believes.

I agree with Chion in part, especially in Dolby's ability to render sounds with exceptional detail, yet I find fault with his analysis of the spatial aspects of the technology. He is correct in noting that Dolby SVA's separation in the flanking speakers is significantly less than in discrete systems, making the sounds seem less distant, but he does not correlate the change in side-channel effect with an increase in surround channel signal. In Dolby Stereo, sounds can move from the side channels into the surrounds or vice-versa. More evocatively, sounds can be made to inhabit the space of the auditorium in counterpoint to other sounds on the screen.

Unlike the use of multichannel surround sound for spectacle, *Days of Heaven* broke with the dominant patterns of effects use in early Dolby Stereo films to experiment with how the surround channels could be used in more subtle ways. Dolby Laboratories vice president Ioan Allen discussed the issues faced by Dolby Stereo in the late 1970s: "The problem we had, if there was a problem, was that we got associated with that kind of film, so science fiction, loud, lots of stereo effects, that's what Dolby meant. It wasn't until probably *Days of Heaven*, Terry Malick's film, that I managed to get across the idea that every film can benefit [from using Dolby Stereo], that quiet, low-level stereo ambiences do as much to help a film as loud spatial sound effects."[7] The surround ambiences help to sell the verisimilitude of the period film by placing the audience into the same acoustic spaces of the characters—letting them listen to the gentle breezes amid the wheat fields at twilight or the rush and whine of the threshing machines at dawn. It was the sound-astute film scholar Vlada Petric who, writing in *Film Quarterly*, identified the main effect of Dolby Stereo's offscreen presence in the film: "The stereo sound emphasizes the spatial aspect of the film image; it works both horizontally and perpendicularly: the audience receives information about the events occurring outside the frame which triggers its imagination constantly. Our visual perception is recurrently challenged by the complex sound track so that we often become aware of the depth of field on the screen by hearing sound which moves from the close foreground and disappears into the far background."[8] Perhaps one of the main ways that Dolby Stereo functioned in *Days of Heaven*, whether 35 mm Dolby SVA or 70 mm, was to create acoustic layers in relation to the image. Instead of drawing attention to the flatness of the screen the effect worked dialectically, with the sound enhancing depth in the image and then, in turn, the image reinforcing the spatialization of the sound.

That is not to say that Dolby Stereo did not have limitations, and filmmakers and technicians alike were struggling to match the capabilities of the technologies to the storytelling needs of their films. In his review of the early Dolby films of the late 1970s, Charles Schreger identified the main conflict as part of an evolution of technology and technique, each one trying to keep up with the advances of the other. Noting the disorienting spatial shifts between the music and dialogue sequences in *The Buddy Holly Story* (Steve Rash, 1978) and *The Deer Hunter* (Michael Cimino, 1978), Schreger had profound praise for Malick's film and averred, "There's no more intelligent use of sound than in *Days of Heaven*—the insects' chirp blending with the machine's whirr, the subtler sounds of breathing and heartbeats—but, with all that care, voices and lip movements are occasionally out of sync; and, when a plane flies overhead, you can hear the shift in sound between speakers on either side behind the screen."[9] Despite Dolby Stereo's capacity for extending the diegetic ambiences into the space of the auditorium, the hyper-realism and spatial dispersion of the sound track often drew attention to misalignments with the image, and the use of moving sound sources were still difficult to position correctly in the mix.

As Chion and Petric noted, Dolby Stereo had the unique ability to construct acoustic layers on a sound track, and to render the sonic and spatial details of each layer with extreme accuracy. Yet the use of Dolby Stereo was never regularly applied in this way because it required that the axis of cinematic action be along a line extending from the screen into the auditorium rather than along the lateral dimensions of the screen. As a result, unlike the example of *Days of Heaven*, many of the earliest Dolby Stereo sound tracks used the surround channel as the location for acoustic effects, rather than for expanding cinematic realism. This was in part due to the preceding strategies adopted from multichannel mixes for four-track 35 mm and six-track 70 mm magnetic format. However, with the need for matrixing and the lack of divergence in 35 mm Dolby Stereo and the bass extension and split surrounds in 70 mm Dolby Stereo, the resultant playback effects were something quite different from the discrete magnetic formats. In order to examine these Dolby Stereo effects, it is necessary to understand the origins of the technologies behind Dolby Stereo and the design, use value, and associated aesthetics of each. And the high-water mark for the conceptual use of Dolby Stereo as part of the storytelling process arrived with *Apocalypse Now* and Walter Murch's concurrent theorization of sound design.

Apocalypse Now (Francis Ford Coppola, 1979) and the Birth of Sound Design

Even though Dolby Stereo achieved earlier successes in films such as *Star Wars* (George Lucas, 1977), *Close Encounters of the Third Kind* (Steven Spielberg, 1978), and *Days of Heaven*, the one film from the 1970s that clearly represented the potential of both creative sound use and Dolby Stereo technology was *Apocalypse Now*. Moreover, it was a film that demonstrated the possibility for a collaborative filmmaking process and how the careful conceptualization of film sound can aid the narrative, and its sound effects and mixing strategies went beyond "added-on effects" to use sound as a powerful tool in the storytelling process. In order to achieve these goals, Francis Ford Coppola and Walter Murch determined that certain technological adaptations were needed, and because of the advanced planning involved in the film's sound they were able to convince both Dolby Laboratories to change their 70 mm Dolby Stereo and theater owners to make the requisite modifications. And, as a result, the aesthetic model developed in the creation of *Apocalypse Now* became the paradigm for sound mixing that would shape the emergence of 5.1 digital sound in the 1990s.

Much of the success behind *Apocalypse Now* is due to Walter Murch's active engagement in theorizing multichannel sound and its use in the film. Taking the screen credit of "sound montage and design" to designate his unique contribution to the film, Murch became involved in the acoustic design of the project when he was brought on as a film editor in 1977. Shortly thereafter he joined a sound team that included mixers Richard Beggs and Mark Berger, sound editors Richard P. Cirincione and Maurice Schell, production sound mixer Nat Boxer, Foley artists Ross Taylor and Kitty Malone, and musicians David Rubinson, Shirley Walker, Mickey Hart, Randy Hansen, and Carmine Coppola in constructing the film's sound. With some scenes utilizing over 200 tracks of sound in the final mix, the Herculean task of creating the sound track was almost as daunting as the two-year production shoot in the jungles of the Philippines. For *Apocalypse Now*, Coppola made three very distinct demands of the sound track: first, that it was in quadraphonic sound; second, there was to be an authenticity of the sound effects and backgrounds; and third, he wanted the sound track "to partake of the psychedelic haze in which the war had been fought."[10] This not only mandated that Murch and the sound team create thousands of new sound effects to establish the verisimilitude of the events, but it also meant that the presentational technology needed to be modified to accommodate the demands of the director. This resulted in the largest sound recording and mixing project for a feature film to date, one where Walter Murch was charged with transforming his concept of sound montage to its next level: sound design.

Murch first developed the idea of sound design as a practice in relation to his contribution to *Apocalypse Now*:

> We felt that, given the equipment that was becoming available in 1968, there was now no reason for the person who came up with the sounds and prepared the tracks not to be able to mix them. The director would then be able to talk to *one person* about the sound of the film the way that he was able to talk to the director of photography about the look of the film. Responsibility for success or failure would lie squarely with that *one person*, and because communication problems would be reduced or eliminated, the chances of success would be increased. . . . On *Apocalypse Now*, however, in addition to picture editing and re-recording, my other task was to develop a design for the use of the film's quadraphonic sound in the three dimensions of the theatre: when should the sound (for dramatic reasons) focus down to a single point; when should it expand across the front of the screen in stereo; and when and how should it use the full dimensionality of the entire theatre? No dramatic film had been released in this format before, so we were moving into uncharted waters.

> I thought I was doing a job similar to the production designer, except I was decorating the walls of the theatre with sound, so I called what I did sound design.[11]

Murch's proposition is intriguing because it represents a form of sound supervision that had never existed. Even during the heyday of the studio system, the function of a studio's sound director was more bureaucratic than aesthetic. Sound directors like Douglas Shearer or Nathan Levinson were less involved with individual films than with the overall maintenance and operation of their respective sound departments. In contrast, Murch's notion that one person could oversee or, better, overhear the entire construction of the sound track was a radical and revolutionary notion. Concurrent was an understanding that new technologies—such as high-speed dubbing units, automated mixing boards, and multitrack tape recorders[12]—made it possible for one person to craft and model the sound of a motion picture.

Although he took the credit of sound designer, Murch never intended to supersede his fellow sound workers; rather, the credit was to differentiate his work from the very specific jobs of his coworkers, letting each individual get full credit for his or her contribution to the film. Effectively, Murch's job as sound designer required that he not only supervise the recording and editing of the sound effects, but that he also participate in re-recording and introduce the idea of sound spatialization in the final version of the film. In the process of creating sound effects Murch was simultaneously interested in maintaining an objective verisimilitude of the sound as well as a conceptual resonance between the sounds and their emotional and dramatic qualities. Emotionally balancing the sounds required occasionally sidestepping questions of realistic reproduction to emphasize the psychoacoustic effect of the sound. As every sound recordist knows, often the sounds heard live on location are not commensurate to their resulting recordings. For example, the recorded sound of a machine gun fired at close range is unimpressively hollow in relation to its destructive power. To heighten the gun's sense of violence and to give the audience a sense of proximity to the weapon, Murch added a recording of brass shells being ejected and hitting the ground. He pointed out that the combined sound wasn't necessarily realistic because in real life the volume of the machine gun would mask the sound of the jangling cartridges; however, by changing the balance of the sounds, "that jangling, metallic sound in combination with the gun added to the emotional fact that you were close to it."[13]

This delicate balance between realistic sound effect use and the incorporation of emotionally resonant sounds, manipulated and artificial, is the trademark of how *Apocalypse Now* does much of its narrative work. Instead of constantly explaining the narrative events to the audience, the film is designed to let them experience the events directly. This is perhaps nowhere better expressed than in the opening sequence of Captain Willard (Martin Sheen) in his Saigon hotel room. The film does not position Willard or his voiceover as a secure narrative anchor for the audience since he regularly slips between objective reality and memories of the war. The fever dream of the introductory sequence sets a pattern for the film's oscillation between concretely realistic representation of sounds and a hallucinatory state brought on by fear and fatigue. As the opening sequence unfolds, the sounds of the battleground transform themselves into the street sounds of Saigon through a stylistic deformation that prevents them from simply becoming an auditory flashback. The central sounds in the sequence are those emanating from a fleet of helicopters seen against a burst of napalm fire. Although the sound effects are realistic in their correlation to the images, as the montage begins the connection of the sounds to a fixed source becomes tenuous when the complex waveforms of the helicopters transform into a series of rhythmic electronic pulses. Despite the obviously fabricated nature of the sound, there is just enough tonal similarity to the initial sound to summon the idea of a helicopter in the mind of the auditor. Yet the transformative shift of the helicopter sound triggers a similar transformation in the image as the rotors of the helicopter are revealed to be the blades of the ceiling fan in Willard's hotel room. This is part of the work devised by Murch and Coppola to keep the audience guessing about the nature of events both seen and heard in the film.

In order to maintain an active sense of audience participation in the story, the film constantly undercuts the security of stable narrative sound usage. Unlike most films, which either cut the sound at the same time as the image or use sound advances to smooth over the spatial transition between scenes, *Apocalypse Now* does just the opposite. The sounds in the film mutate and transform, making an implicit connection between one scene and the next but never offering the stable acoustic ground of classical Hollywood film sound construction, and challenging the simple assumption that a sound should always remain attached to its source. As an example, when Willard comes out of his stupor, the film presents the spectator with an image of the ceiling fan revolving at the same rate as the rhythmic pulses. What had previously been heard as the sound of a helicopter now is heard as the rhythmic sweep of the ceiling fan. Yet as Willard rises from his bed and observes the bustling Saigon street from his hotel room window, the audience simultaneously hears the simple electronic sound change into a real recording of a helicopter. In this way the distortion of the sound effect is verified as being the sound of an actual helicopter outside of the hotel. Again, this is only speculation, because when Willard peers out his window there is no image of a helicopter or any indication of its presence outside of the frame. The stability of identifiable sound effects and their ontological link to diegetic objects is regularly undercut throughout the film to achieve a higher level of interplay between sound and story.

By contrast, the only stable acoustic space presented in the film is that of Willard's internal monologues. Instead of simply adding narrative information as a form of external commentary, Willard's internal voice functions as a direct perceptual link to the events that are unfolding. Different from the omniscient narration of many Hollywood films, Willard's impressionistic narration is temporally linked to the events taking place and his reaction to them.[14] By combining a shifting framework of diegetic sounds with a stable internal narration, the film effectively inverts several of the assumptions regarding normal film sound. The result is that a constant flux of sound influences the auditor directly through the "psychedelic haze" of the sound track.

In addition to the complex relationship between internal and external sounds, the film also structures an interesting distinction between onscreen and offscreen sounds. *Apocalypse Now* generally refuses to show the source of sounds, and in so doing creates a strong sense of ambience while also adding an element of danger to the offscreen space. Insect-like buzzing, whistles, hums, cracks, booms, swishes, and dozens of other sounds populate the offscreen sound track and remain resolutely acousmatic. Unlike the perceptual "call-and-response" game of *Star Wars*, where sounds are regularly introduced on the sound track, pointed out in dialogue, and visually sourced as confirmation, the mystery of the offscreen sounds is never revealed or explained in *Apocalypse Now*. Instead, offscreen space functions as a liminal zone, concealed from optical inspection and constantly eluding attempts to unveil its mystery. Like Willard and his companions on the PBR boat, the audience can only see what is on the shoreline, but the sounds of the jungle come forth to penetrate their perception and work insidiously to plunge them deeper into the story along with the characters. The offscreen sounds in *Apocalypse Now* manage to achieve the status of pure aural objects. Unrestricted by visual signification or narrative explanation, they operate as free-floating signifiers designating any number of possible sources but refusing to be fixed to any single source.[15]

An example of this is the "nervous insect" sound heard as Willard and Chef (Frederic Forrest) leave the PBR boat and move through the jungle. Although the sound can be described as "insect-like" because of its high-pitched throbbing and the way it flits from speaker to speaker, the sound itself is unmistakably electronic and does not correspond to any known insect. However, by positioning the sound halfway between an actual insect sound and a purely conceptual one, the film engages the audience in a guessing game of decoding and deciphering the function of the sound. In this way it induces the requisite nervousness and attentiveness in the audience that matches the characters' heightened sense of danger. When the electronic sounds reduce to nothing before the appearance of a lurking tiger, the audience is already in a fully heightened state of awareness that makes the attack even more startling.

As with *THX 1138* (George Lucas, 1971) and *The Conversation* (Francis Ford Coppola, 1974), Murch carefully manipulated the voices in *Apocalypse Now* to give them secondary functions as effects. Not only were the voices used to convey narrative and emotional information about the characters, but by adding spatial characteristics and degrading the sound of the voices in certain circumstances they were placed firmly within the space of the diegesis. This is partially because of Murch's interest in maintaining an absolute sense of realism when re-creating the environment of Vietnam, but also due to the exigencies of filmmaking. Because the location dialogue was recorded under some of the most extreme circumstances imaginable, very little of it was usable for the final film. This meant that a large proportion of the film needed to be looped in order to restore most of the dialogue to a level of intelligibility. Because, as in *Star Wars*, the quality of the Dolby Stereo reproduction technology was such that it would reveal several of the shifts in background sound between location and looped dialogue, it was decided that virtually all dialogue should be replaced in the film. Of course, the replacement of dialogue comes with its own problems, particularly the fact that nearly all background sound is lost in the process and the voices sound unnaturally clear when heard in the film. To avoid this effect, Murch and his sound team had the actors reenact their movements in the recording studio, thus constructing a system known as "real ADR" to replicate the effect of battle in the ADR process. Effectively a combination of Foley, group walla, and ADR procedures, "real ADR" was used both to augment the performances of the actors during looping and to provide a spatialized sense of the sounds for the film. As Murch pointed out:

> The problem in a conventional "loop group" session is people are just standing in a room, and you don't get the sound of people actually talking while they're running. . . . So you get these people moving in space, really exerting energy, really yelling, and always shifting in relationship to the microphone. . . . So the helicopter attack sequence is a sandwich of four different kinds of dialogue. There is some production dialogue, Duvall mostly. There is this pilot chatter. There are the kids on the ground running, and there is "real" ADR dialogue.[16]

The use of the real ADR allowed Murch to incorporate the shifting dynamics and spatialization of the characters into the final film, thereby creating a vastly improved sense of realism over just looped dialogue.

Another way that Murch manipulated the dialogue in the film was through the direct recording of the lines through transmission devices such as radios and com-links. In doing so he was able to shape the acoustic characteristics of the sound in the looping process rather than in the final mix. This also added a level of verisimilitude to the performances that allowed the actors to act, and react, in a way that was appropriate. When Robert Duvall was preparing for the scene of the helicopter attack, Murch kept urging him to increase the loudness of the voice. Although Duvall was familiar with the proper voice levels for the highly sensitive microphones generally used on a film, he was not used to the idea of having to speak his lines for the helicopter's communication equipment.[17] By recording the actor's lines through the com-link microphone, and forcing the actor to shout the lines to be heard over the din of the helicopter, Murch was adapting the technology and his methodology to suit the needs of the story. Moreover, this was a different approach than the manipulation of dialogue in the final mix to accommodate narrative intelligibility. Murch was able to use the distortion from the radio sets to enhance the quality of the dialogue rather than to obscure it. The end result is both a powerful performance on the part of the actors and a navigable field of interactive dialogue and sound effects.

The final form of dialogue manipulation in the film refers to the very intimate sound of Willard's internal narration. Unlike most of the dialogue, which was generally restricted to the center channel of the screen, Willard's voiceover was fed to all three of the front speakers equally. Walter Murch came across the idea of Willard's internal narration, or "head voice" as he called it, after working with mixers Bill Rowe and Gerry Humphries on Fred Zinnemann's *Julia* (1977).[18] The idea for the close-miked vocal recording was originally developed by British sound editor Les Hodgson to differentiate Ishmael's internal narration from the regular dialogue in John Huston's *Moby Dick* (1956). Although the idea of close-miked narration was certainly not a new cinematic practice, the radical differentiation between the normal speaking voice and the near whispered narratorial voice of a character was new to American films. This use resuscitates the British mixing strategies that developed in the 1950s to emphasize the sound of "up-front dialogue" even when a character is offscreen or speaking at a lower volume. Early examples come from the films of Richard Lester, where despite the large number of persons talking in a location the lines of dialogue are heard clearly at approximately the same volume level despite their differing locations. This aesthetic of close miking, obscured by a decade of Hollywood sound and image scale matching, was rejuvenated in the narration for *Julia* and subsequently adapted as the distinguishing mark of Willard's subjectivity in the *Apocalypse Now*. This allowed Murch and Coppola to differentiate between the objective realism of the spoken dialogue and the subjective impressions of Willard's monologue. As well, it created a secondary level of narrative information in the film, one that could be deployed simultaneously with the diegetic sounds without becoming lost in the mix.

The process of editing and mixing the sounds in the film was also carefully considered in advance to provide as much narrative information as possible through the sound track. The fact that all the sound effects in the film were recorded in stereo allowed for the creation of a natural form of divergence in the 35 mm Dolby Stereo mix as well as in the 70 mm version. In order to compile such a complicated sound track, two separate studios were constructed for Coppola: one to edit the film and another to mix it. This required a large-scale effort to incorporate as many technological advances as possible to aid and facilitate the process. To ensure that the quality of the sound would remain high throughout the post-production process, all the tracks received dbx noise reduction to provide a 20 dB increase in the signal-to-noise ratio.[19] The editing itself was done on an automated MCI mixing console with a computerized memory assist and specially modified to allow an output of six discrete channels. The mixing console was connected to two 24-track recording decks and high-speed servo dubbers to accommodate Coppola's demands of quadraphonic sound.[20] Nearly all the technology used in the post-production process was drawn from the music recording industry instead of the film industry. Not only was this because the film industry had never embarked on such a complex sound track before, but also because the majority of the sound team had come from the music industry before moving into film.[21] According to Omni Zoetrope engineer Terry Delsing: "All this technology suddenly fell upon the film world, which had a narrow tunnel-vision approach to sound. The median age of sound mixers in Los Angeles, the 'A' roster, was 50 to 60 years old. Then all of us nutcake, burned out hippie, rock 'n' roll rejects were put into these situations and told to mix a film."[22] Although Delsing downplayed the talent of the sound team, he offered an important point about the changing regimes both inside and outside Hollywood. As the film industry matured, so too did its employees. With the breakdown of the apprentice system in the late 1960s, the only place from which to draw new talent was the music recording industry. And by carefully choosing his collaborators and their background, Coppola was able to inject a vitality into the film that was impossible to achieve in Hollywood.

The strategy that Murch developed for editing and mixing the film was one of "overlapping aural imagery" where the sounds worked together to achieve a greater level of meaning in the film.[23] Generally when a large number of sounds are edited and mixed together in a film, as with *Star Wars*, it can result in a cacophony that destroys the individual distinctness of each sound in the mix. Murch was determined to avoid this at all costs because the sound track needed to achieve a balance between judicious sound use and narrative fluidity. He articulated this desire in the following way: "The guidewords I keep engraved somewhere in the back of my mind are *dense clarity*, or clear density. What I want is something that gives the impression of that density of reality, yet looking into it, it is very clear. It does not fuse into a block, or change into something other than its parts. At the same time I'm interested in extreme clarity, but I want that clarity to have a weight to it, and that is only achieved by a great number of separate parts set at the right balance."[24] To achieve this balance between a large number of separate sounds and the clarity of the narrative, Murch edited sounds together in a way that emulated the hyperselectivity of the brain in regard to hearing. Hyperselectivity is the way that our brains filter out unwanted noises to concentrate on distinct sounds. The most common example is when several people are speaking in a room and we have to concentrate on what a single person is saying, a function Colin Cherry has called the "cocktail party effect."[25] Murch was more interested in exploiting the way in which this form of acoustic filtering could be applied to the editing and mixing process to heighten the dramatic aspects of the film. As he noted, "The sound is what's in your mind, not what is on a piece of magnetic tape."[26] Although he uses the idea of hyperselectivity throughout the film, two very clear examples demonstrate the effectiveness of the device.

The first occurs when Willard and the PBR crew arrive at the Do Lung Bridge near the border of Cambodia. The scene itself is shot with extra anamorphic filters on the camera to slightly distort the image and convey the sheer insanity of a bridge that gets destroyed every day and rebuilt every night (see figure 24). While searching for the commanding officer to retrieve information about his mission, Willard stumbles upon two soldiers trying to kill a Viet Cong sniper at the perimeter of the battleground. Although the Viet Cong soldier cannot be seen, he is heard repeatedly taunting the Americans amidst the din of rocket fire, machine gun fire, and arc welders from the bridge repairs. To solve the situation the soldiers call on the services of "Roach" (Herb Rice), who proceeds to aim his grenade-launcher at the Viet Cong soldier by listening attentively to the location of the voice. To emphasize the character's perceptual shift in relation to his concentration, Murch edited out certain sounds to emulate the effect of hyperselectivity. He explains: "There's a transposition with sounds of the battle being replaced by bridge building because I wanted you to hear his breathing, the jingling of his beads, *to hear with his mind* as he listens to the Viet Cong in the distance. He's aiming his gun not by seeing him but by sound. Explosions around him would be distracting under these circumstances, so he's shut them out."[27]

A similar use of selective sounds occurs during the following scene and the flare attack on the PBR boat. In the mail parcel at the Do Lung Bridge was a cassette-recorded letter from the character Clean's mother. While Clean (Laurence Fishburne) is listening to the recording the ship comes under attack from hundreds of flares shot from the shoreline. At the moment when the attack starts, the sound track is filled with diegetic sounds from the boat, nondiegetic

FIGURE 24. The "psychedelic haze" of the Vietnam War reaches its apogee during the Do Lung Bridge sequence in *Apocalypse Now* (Francis Ford Coppola, 1979).

score music, and the voice of Clean's mother on the tape. During the attack, the balance of the sounds shifts to emphasize the battle sounds by eliminating both the music and Clean's mother. After the attack, when they notice that Clean has been shot, the score music returns followed shortly by the voice on the cassette. As the crew gathers around Clean's body, all the sounds of the boat gradually fade out, leaving just the score music, Chef's quiet sobbing, and the voice of Clean's mother saying "stay out of the way of the bullets." Although the audience does not consciously note the shift in sounds or the fact that certain sounds leave and return, the psychological effect is extremely powerful due to its emulation of the way that the soldiers perceive their environment.

The verisimilitude of the sound effects and the "psychedelic haze" was also carefully constructed through a judicious use of the final multichannel mix. Although films had been mixed in multichannel stereo since the 1950s, the creation and theorization of a quintaphonic mix is a unique development in the evolution of American film sound. According to Michael Rivlin of *Millimeter* magazine, "In order to create an additional discrete channel of sound, Murch reasoned that because low frequency sounds were not capable of being perceived as distinct left and right point sources, stereo separation of low frequency sounds—as it existed in the conventional [70 mm Dolby Stereo] six-track mode—was being wasted. Instead, Zoetrope fed a single mono source to the two low frequency speakers in front and used the extra track to carry a discrete signal to a seventh speaker in the rear to create true stereo surround."[28] John Mosely's five-channel "Quintaphonic" setup for *Tommy* was the initial inspiration for the quadraphonics, yet there were some serious drawbacks to the equal weighting of each of its channels in its theatrical application.[29] A careful modification of the quadraphonic concept was necessary to make it function according to Coppola's demands and to adhere to the available technology. Specifically, although quadraphonic sound dispersion was expected, there needed to be a way to weight the sounds toward the screen channels and avoid deafening those in proximity to the rear channels. In addition, the 70 mm Dolby Stereo technology needed to be modified to provide two discrete sound channels while also providing low-frequency reinforcement.

What follows is Walter Murch's recollection of the development of the multichannel system and its application to the final mixing of *Apocalypse Now*:

> It unquestionably originated with Francis's idea to have the film be quadraphonic. . . . I was [in 1976] in England mixing *Julia* with Bill Rowe, who had mixed *Tommy*. So I was able to question him about everything, and Mosely, and that system, and observations about how it worked and didn't work. . . . *Apocalypse Now* is obviously a very different animal than *Tommy*. It's not a musical; it's a dramatic action, character, mood film with a very complex, dynamic, changing soundtrack. . . . So I began by thinking, all right, let's make it five channels. We'll have quadraphonics, which is to say a source in each corner of the room. . . . But we'll have a center channel. So this is essentially Mosely's quintaphonic approach.[30]

Although Bill Rowe did not use the center channel in Mosely's Quintaphonics exclusively for voice, it did offer Murch the stability he needed to have *Apocalypse Now* shift between mono, stereo, and quadraphonic sound. This capability of shifting the sound balance from quadraphonic to screen stereo to mono was made possible with the inclusion of a joystick control to channel sounds anywhere in the 360° soundfield.[31] By using the joystick control it became possible to place and move sound effects anywhere within the soundfield, and it also enabled the sound team to expand or collapse the soundfield from quadraphonic to monophonic when needed.

Throughout the film the surround channels and potential quadraphonic presentation are used judiciously. Although the surround channels are often used to spectacular effect, especially during the previously mentioned scenes of Chef and Willard in the jungle or the flare attack on the boat, most of the time the sounds of the film are restricted to the three speakers behind the screen. This is for two basic reasons, one purely aesthetic and the other purely technical. First, because audiences are conditioned to expect all narratively significant sounds to come from the space of the screen, Murch limited his use of the surround channels to avoid shocking or overwhelming the audience with their presence. The main design flaw in Mosely's Quintaphonics was the equal weighting of the speakers and the fact that the surround channels were always engaged. This placed an enormous strain on the audience and particularly on anyone who was unfortunate enough to be seated near one of the rear speakers. By prudently activating the surrounds only when the sounds were tied to significant narrative moments, Murch was able to surround the audience with sound while letting the surround channels do some of the narrative work. A prime example of this is the arrival of Willard and the PBR boat at the beachfront to meet the helicopter cavalry. As the boat approaches the beach from the distance, the sounds of the ensuing battle gradually move from mono to stereo across the three front speakers. Yet when Willard and the crew land and move further ashore, the sounds of battle move into the surround speakers to emulate the events around them. But what is more important at this point is that as the crew moves further into the scene, the same sounds of battle surround the audience, thereby pulling them into the diegesis. This effect is quite powerful and only used a few times throughout the film so as not to wear out the audience. However, it needs to be noted that this use of the surround channels was only constructed for use with the 70 mm Dolby Stereo mix. The second reason why most of the sounds were kept on the three channels of the behind-the-screen speakers is because the 35 mm Dolby SVA version was mixed to have *absolutely no surround information.* Murch reasoned that the complicated psychoacoustics of the surround channels demanded a great deal of control over the reproduction equipment. Because Dolby SVA was not as flexible as the discrete channels of 70 mm, and since there was no standardization in most theaters for 35 mm surround channel levels, the idea of using the surround channels was abandoned altogether.[32]

Coda: The Legacy of Dolby Stereo and Sound Design in the 1980s and 1990s

Hence, in considering *Apocalypse Now* as the zenith of creative sound work in the 1970s, the fact remains that the bulk of its creativity lay in the technological choices that informed its aesthetic. The use of a modified form of 70 mm Dolby Stereo with bi-amped surround channels allowed the sound team to develop a sound aesthetic that met the demands of the director and the story being told. The end result was a film that mobilized sound not simply to augment the visual narrative, but to be an equal partner in the process of storytelling. Never did the sound team feel that they were supposed to foreground their sound work or demonstrate the potential application of the sound technology to other films. If anything, they were extremely understated in the assessment of their contribution, preferring to remain unnoticed in the process. As Walter Murch explained to Charles Schreger: "The most we (sound people) can hope for—and I mean in a very positive sense—is that audiences will feel incredible emotion and tension or surprise or relief or whatever—and ascribe it not to the sound, but to the actor or to the scene or the photography. For the most part, it's much better for us to be able to work on people without them really knowing it."[33] Through the stereophonic recording of original sound effects, the editing of sounds to emulate the psychoacoustics of perception, and the subtle utilization of the quadraphonic sound field, the elements not only shaped the sound aesthetics of *Apocalypse Now*, but also offered a theoretical template for the future of film sound with the advent of 5.1 digital sound tracks. Finally, the most important element in the success of the sound of *Apocalypse Now* was the willingness of the director to place full confidence in the sound team and the creative freedom afforded to them.[34]

It was extremely rare for one person to function as a supervising sound editor and re-recording mixer on a film, and, indeed, there were additional interdictions preventing an individual from working in both the production and post-production phases. The difficulty for sound design to reach its potential in the 1980s and 1990s was certainly not due to the validity of Murch's initial intention. Instead, a number of competing factors have troubled the history of the role, splitting it between the positions of supervising sound editor, re-recording mixer, and sound effects creator. As Randy Thom noted twenty years after the introduction of the term "sound designer": "Walter Murch's dream of someone with a 'sound mind' guiding the use of audio throughout the project is taken no more seriously now than it was a decade or two ago. Lots of lip service is paid to the value of sound, and the now near-meaningless term 'Sound Designer' proliferates. . . . But the sad fact is that those of us interested in sound for its potential as a storytelling partner have no more control over (or even influence on) that process than we ever had."[35]

Walter Murch's role remains unique in the history of film sound, and his distinctive position was enabled due to his location at the periphery of the film industry. Because he was able to work out of San Francisco instead of the strongly union-controlled studios in Los Angeles, he had a measure of freedom unavailable to most sound workers. George Lucas expressed this ethos best, noting that in Los Angeles, "you're just asking for trouble, because you're trying to change a system that will never change. [In San Francisco] we don't change a system, because there is no system."[36] Also, Murch was fortunate enough to work with directors like Coppola and Lucas who were willing to temporarily unhinge the narrative demands of intelligibility and plot motivation that had become an ideological core of the Hollywood production process. And even though there have been many who approach the role of a sound designer by supervising the creation of the sound track, it wasn't until the late 1990s that other individuals have been able to realize the ideas behind Murch's conception.

Larry Blake, an engineer and writer on film sound who went on to become a major supervising sound editor and re-recording mixer, summed up the state of film sound in the early 1980s as follows:

> It is a sad but true fact that the majority of soundtracks produced in Hollywood are not the end result of one person's taste, constantly refining ideas from the beginning of pre-production, as is the case with the cinematographer's contribution to the image. Instead, most soundtracks are slapped together by a crew of sound editors after a fine (final) picture cut is made. Given only a few weeks to prepare the cut sound elements for final dubbing, each sound editor will cut only a reel or two (out of a total between 8 and 12 used in an average film), and hence is unable to stand back and look at the overview. Even though there is always a supervising sound editor, that person, too, is handicapped by the time constraints.[37]

Blake's statement indicates that in the majority of cases both the labor system and demands of production made it impossible for the role of a sound designer to emerge.

By the end of the 1970s the widespread acceptance of Dolby Stereo as the preferred presentational format was actively transforming the landscape of American film sound. In 1980, over 1,200 theaters in the United States and 500 more worldwide were equipped for Dolby Stereo playback, with the vast majority of these being 35mm Dolby SVA.[38] Within a year, the number of domestic theaters with Dolby Stereo grew to 2,000 while new SVA prints were outpacing standard monophonic by nearly two to one.[39] As the decade progressed, Dolby Stereo's presence in the marketplace grew exponentially. May 1984 saw the release of the 500th film encoded in Dolby Stereo, and by November 1986 over 1,000 films had been released in the format.[40] By the end of the decade Dolby Laboratories had already introduced their Spectral Recording format, Dolby SR, which further improved the noise reduction capabilities and dynamic range of the optical sound track. Dolby SR represented the final step in squeezing all of the potential sound range out of an optical sound track, yet it did not mark any appreciable change in the basic technology and its commensurate aesthetic properties. Therefore, the "Dolby decade" of the 1980s was one of rapid industrial transformation with minimal aesthetic shift.

The double aesthetic of Dolby Stereo, promoted through the discrete channel 70mm system but marketed as the 35mm Dolby SVA format, continued throughout much of the 1980s as the dominant trend of film sound. The 70mm system was the format of choice for most first-run films during the decade, and many of the films continued to propagate the same stylistic choices as *Star Wars* and *Close Encounters of the Third Kind*. By foregrounding sourced sound effects and score music, these films continued the trend of "louder is better" in films such as *The Empire Strikes Back* (Irvin Kershner, 1980), *Superman II* (Richard Lester, 1980), *ET: The Extra-Terrestrial* (Steven Spielberg, 1982), and the Indiana Jones franchise. However, in the absence of a definite aesthetic design for the Dolby SVA format, several filmmakers began to explore the possibilities of the technology. *Invasion of the Body Snatchers* (Philip Kaufman, 1978), *The Black Stallion*, *Alien* (Ridley Scott, 1979), *The Elephant Man*, *Blade Runner* (Ridley Scott, 1982), and *Dune* (David Lynch, 1984) all developed stylistic approaches that exploited the technical attributes of the Dolby SVA system to envelop the audience in ambient sound. The resultant aesthetic played into the narratives of each of these films to create a structural relationship between the film's sound and its story.

Arguably the film that solidified the conceptual resonance between sound and storytelling is Martin Scorsese's *Raging Bull* (1980). For the film Scorsese worked with sound effects specialist Frank Warner, who developed the signature sounds during the boxing sequences. Specifically, Warner supervised what he referred to as the "point-of-view special effects" to develop the subjective effect that draws the audience into the perception of the main character, Jake LaMotta (Robert De Niro), whenever he is in the ring.[41] During these scenes time becomes elastic and audiovisual events expand or contract according to the dynamics of the story. According to Gianluca Sergi:

> The non-literal approach to sound in the movie, especially during the fight sequences, has a dramatic effect on the overall feel of the movie and the way it impacts on audiences. In particular, Warner and Scorsese's choice of designing the breathing of the boxers as a combination of animal sounds played back at different speeds and mixed with other sounds conveys an eerie, unsettling quality to the fight scenes. The mixing of those sounds with the aggressive sounds of the camera flashes documenting the fight creates very effective contrasts between the very personal world the boxers inhabit when in the ring and the outside world's sadistic desire to witness their public humiliation. . . . The sound orchestration in *Raging Bull* is as daring as it is inventive and effective.[42]

In addition to the use of sound to convey a sense of character subjectivity and to render the violence of the boxing matches, Dolby Stereo technologies also expanded the world of the story from the screen into the space of the auditorium. In both the 35mm Dolby SVA and the six-channel 70mm release, the fight sequences progress from a documentary style realism that anchors the punches and crowd noises in the screen channels to a hyperbolic effect as the sounds migrate into all the channels in the auditorium. During LaMotta's initial fight with Sugar Ray Robinson (Johnny Barnes), his perennial opponent,

FIGURE 25. Martin Scorsese used the multichannel screen speakers and surround channels of Dolby Stereo to emulate the subjective experience of Jake LaMotta (Robert De Niro) during the boxing sequences in *Raging Bull* (1980).

crowd sounds expand into the surround channels along with the whooshes and whirs added to the camera pans and the cracks of the flashbulbs. With each subsequent Robinson match the sound shifts from objective to subjective as more abstract sounds are introduced. Drones, animal cries, an electronically altered timpani, and even thunderclaps are used to emulate the haze and confusion of the bout, and the sounds move into the surrounds while the punches and radio announcer voiceover remain in the screen channels. As Jake falters in the third fight against Robinson, the exaggerated soundscape works to emulate his fallibility (see figure 25). After the tropes of the fight sounds are introduced, the audience experiences more of the dramatic flow of the fights through the sound effects work rather than through the ringside announcers' play-by-plays. The combination of the integrated sound design and the use of Dolby Stereo let Scorsese craft his intimate portrait of LaMotta's rise and fall.

Scorsese's collaboration with Warner, which began on *Taxi Driver* (1976), carries with it all the earmarks of sound design as theorized by Walter Murch. Commenting on this point, supervising sound editor Mark Mangini wrote:

> One must remember that sound design is, first and foremost, a forum of ideas, not technology. . . . I remember years ago being humbled by the work of Frank Warner. At a time when I was feeling pretty smug about using state-of-the-art tape recorders and samplers, I heard Frank's work on *Close Encounters of the Third Kind* and *Raging Bull*. He was still using good old 35 mm mag, fundamental tape manipulation techniques and sounds

from a venerable library to turn out the most striking soundtracks around. Frank was a genius in his day, because he understood drama and pacing, mood and color, not sample rates or frequency response or bandwidth.[43]

Mangini's comment contains a hidden point in the fact that cinema sound technology was undergoing a major sea change in the late 1970s and early 1980s, not only because of Dolby Stereo but also because of new multichannel sound editing techniques and the advent of digital recording and reproduction.[44] With the subsequent changes in post-production audio technologies came a reliance on preexisting sound aesthetics while the practitioners learned the capabilities of the new technologies.

As an object lesson, *Raging Bull* can be seen and heard as embracing both the advantages of Dolby Stereo's enhanced dynamic and frequency ranges, as well as its sound spatialization, to let sound become an equal partner in the storytelling process. Indeed, Dolby technologies made it possible for filmmakers and film sound practitioners to explore new and prior modes of audiovisual representation. As Michel Chion pointed out, "The loud pops of midcentury flash cameras in the final boxing match in Scorsese's *Raging Bull* (1980) reinforce the idea of vertical montage points (matching visual with audio) and thus participate in a return in the 1980s to the sensorial and to the foregrounding of montage, two features commonly associated with the flamboyant period of silent film."[45] Yet, as was often the case, the conflicts among established practices, labor dynamics, studio pressures, and technological demands restricted the speed with which aesthetic advances kept up with technology. It is important to remember that aside from the impressive use of Dolby Stereo's spatialization during the fight sequences in *Raging Bull*, the majority of the film was presented in mono,[46] with all dialogue and sound effects channeled to the center speaker, in a style that emulates its 1940s and early 1950s black-and-white visual aesthetic. Indeed, most films from the early 1980s still clung to the established aesthetics of the late studio era and few chances were taken to exploit the potential of the Dolby Stereo technologies. As with the case of the studio-imposed voiceover narration in *Blade Runner*, or the redundancy of sound and image relations in films like *Blow Out* (Brian De Palma, 1981),[47] the possibility of developing new audiovisual aesthetics was not readily embraced by the industry at large, and early experiments gradually faded out in lieu of "filling" the sound tracks with a heavy-handed use of sound and music. It was not until later in the 1980s and the early 1990s that directors worked to change this dominant paradigm through an active exploration of Dolby Stereo and its aesthetic potential.[48]

If there is one distinct result of the hegemony of Dolby Stereo in the 1980s, it is the obfuscation of many of the different aesthetic approaches to film sound in the previous decade. Nearly all the techniques and approaches that were developed in the 1970s either lost favor in the 1980s or were incommensurate with the techniques of Dolby Stereo. One of the prime examples of this is the use of Robert Altman's multichannel recording apparatus and the introduction of Dolby Stereo. Developed roughly at the same time, Dolby Stereo and Altman's close radio miking and overlapping vocal aesthetic were ultimately incommensurate. Because Altman's sound mix was carefully selected to have no spatial characteristics and a complete lack of reverberation on the voices, it required carefully controlled recording and mixing strategies to ensure intelligibility while also letting the voices compete for attention in the mix. These techniques were entirely at odds with the strategies of Dolby Stereo mixing practices and would result in the collapse of all voices to mono and a sheer unnavigable cacophony in the final mix. In an attempt to circumvent the growing trend toward Dolby Stereo, Altman convinced Paramount Pictures to develop the Parasound multichannel optical system for use on *Popeye* (1980). Based on principles similar to Mosely's Kintek, the sound system failed to provide the multichannel control that Altman demanded. As a result, Altman was forced to change his strategies to no longer include overlapping voices in his films of the 1980s, but to develop a system of concatenated dialogue, where one person starts speaking as soon as another one stops. Though this gave the impression of nonstop dialogue in the picture, it sacrificed the potential of the democratic sound used in Altman's previous pictures.

The result of Dolby Stereo's ubiquitous presence in 1980s cinema sound was that many of the aesthetic experiments of the 1970s either were lost or lay dormant until they were picked up by innovative filmmakers decades later. The democratically organized sound team, which had opened up so much potential for creative sound use and construction in the 1970s, was decimated in the 1980s by the inflexible strictures of Dolby Stereo mixing practices. No longer was it possible for individuals to straddle the gap between production and post-production to help guide the film's sound throughout the filmmaking process. Instead, the regulations governing sound mixing insisted on a standardized approach to both dialogue recording and sound effect construction to ensure a stable sense of sound use in all Dolby Stereo films. Moreover, the hierarchical divisions of labor continued to be maintained in Los Angeles through the strength of the labor organizations, making it nearly impossible for either the New York or San Francisco approaches to take root in Hollywood. Most importantly, this did not leave much room for true sound theorists like Walter Murch, who shifted his focus from film sound to picture editing in the 1980s.

The creative legacy of 1970s film sound continues, however, as a potential model for rediscovery in the current era of 5.1 digital sound. In the 1990s the strategies of sound recording and mixing started to fragment in the transition to the discrete channel approach of digital sound. The return to a discrete channel format, in contrast to matrixed Dolby SVA, meant that several of the 1970s sound experiments were once again viable prospects. Because all the elements on the sound track needed to be mixed into the stereo field, no one type of sound would necessarily take precedence in the final mix. This reopened the possibility for sound effects and ambience becoming as important as dialogue in the storytelling potential. As well, it accommodated the creative interaction between source and score musics to blur the boundaries between the diegetic and nondiegetic. But most importantly, the elimination of an overriding and technologically derived aesthetic in 5.1 digital sound brought about a complete rethinking of the interplay between the sound track and the visual apparatus. The sound track's ability to take over the function of the establishing shot or shot/reverse shot gave rise to an ontological breakdown in the status of the image and a newfound awareness of the constructed nature of cinema. As a result, directors today have more freedom than ever to reconceptualize the interplay between sound, image, and story and to rewrite the audiovisual contract.

NOTES

CHAPTER 1 INTRODUCTION

1. Robert Phillip Kolker, *A Cinema of Loneliness: Penn, Kubrick, Scorsese, Spielberg, Altman*, 2nd ed. (New York: Oxford University Press, 1988), 9.
2. Rick Altman, *Silent Film Sound* (New York: Columbia University Press, 2004), 15–21.
3. See Kolker, *A Cinema of Loneliness*, and Robin Wood, *Hollywood from Vietnam to Reagan* (New York: Columbia University Press, 1986).
4. Rick Altman, "Introduction: Sound's Dark Corners," in *Sound Theory/Sound Practice* (New York: AFI/Routledge, 1992), 171–177.
5. While Robert Altman's films hold a plurality of voices, opening up a variety of audio "pathways" through the narrative, they never dwindle into complete cacophony, due to the careful mixing practices of his sound and editing teams. Often there is one central conversation heard in each scene or a central character that serves as a point for audience identification and following. For more on this see Rick Altman, "24-Track Narrative? Robert Altman's *Nashville*," *Cinéma(s)* 1, no. 3 (1991): 102–125, and Jay Beck, "The Democratic Voice—Robert Altman's Sound Aesthetics in the 1970s," in *A Companion to Robert Altman*, ed. Adrian Danks (Hoboken, NJ: Wiley-Blackwell, 2015), 184–209.
6. In keeping with this book's focus, part of my project is to point out and challenge the hegemony of visual metaphors that dominate cinematic discourse. It is nearly impossible to write about film sound without the use of visual metaphors or figures of speech, and even though this project does not seek to overturn the dominance of the visual in film studies, it is designed to foster a new sensitivity for the acoustic dimension of cinema.
7. See Thomas Elsaesser, "Notes on the Unmotivated Hero—The Pathos of Failure: American Films in the 70's," *Monogram* no. 6 (October 1975): 13–19.
8. Douglas Gomery, "The American Film Industry of the 1970's," *Wide Angle* 5, no. 4 (1983): 52.
9. Kristin Thompson, *Storytelling in the New Hollywood* (Cambridge, MA: Harvard University Press, 1999), 4; my emphasis.

CHAPTER 2 THE BRITISH INVASION

1. See John Belton, "1950s Magnetic Sound: The Frozen Revolution," in *Sound Theory/ Sound Practice*, ed. Rick Altman (New York: Routledge, 1992), 154–167.
2. Paul Monaco, *The Sixties: 1960–1969* (Berkeley: University of California Press, 2003), 102.

3. Aubrey Soloman, *Twentieth Century Fox: A Corporate and Financial History* (Metuchen, NJ: Scarecrow Press, 1988), 162.
4. See Andrew Sarris, "Sixties Cinema: Zoomshots, Jumpcuts, Freeze Frames, and Girls, Girls, Girls!," *Rolling Stone* no. 148 (22 November 1973).
5. See David Cook, *Lost Illusions: American Cinema in the Shadow of Watergate and Vietnam: 1970–1979* (Berkeley: University of California Press, 2000); Peter Lev, *American Films of the '70s: Conflicting Visions* (Austin: University of Texas Press, 2000); and Ryan Gibley, *It Don't Worry Me: The Revolutionary American Films of the Seventies* (New York: Faber and Faber, 2004).
6. Geoffrey Nowell-Smith, *Making Waves: New Cinemas of the 1960s* (New York: Continuum, 2008), 124.
7. Gene Phillips, "John Schlesinger: Social Realist," *Film Comment* 5, no. 4 (Winter 1969): 59.
8. Joseph Gelmis, *The Film Director as Superstar* (Garden City, NY: Doubleday, 1970), 240.
9. Stephen Watts, "The Beatles' 'Hard Day's Night,'" *New York Times* (26 April 1964): X13.
10. Manny Farber, *Negative Space: Manny Farber on the Movies*, expanded ed. (New York: Da Capo, 1998), 160.
11. Haskell Wexler is a particularly important figure because he included several of the film's audiovisual strategies in *Medium Cool*; Hal Ashby's editing on *The Thomas Crown Affair* (Norman Jewison, 1968) was influenced by Richardson's willingness to separate sound and image in the editing room. *The Loved One*'s use of familiar music as score ("America the Beautiful," the "Liebestod" from *Tristan und Isolde*) may have also influenced Ashby's musical choices in *The Landlord* (1970) and *Harold and Maude (1971)*.
12. In the documentary that accompanies the film's DVD release in the United States, actor Robert Morse comments that all of his lines were looped after the initial shooting. Some dialogue, however, is recorded on location (most notably the parts played by Jonathan Winters and Dana Andrews) and the difference in background tone is striking.
13. Philip French, "Point Blank," *Sight & Sound* 37, no. 2 (Spring 1968): 98.
14. David Thompson, "As I Lay Dying," *Sight & Sound*, ser. 2, vol. 8, no. 6 (June 1998): 17.
15. In his review, Andrew Sarris compares Boorman to Alain Renais when he refers to *Point Blank* as "Last Year at Alcatraz." *Confessions of a Cultist: On the Cinema, 1955–1969* (New York: Simon and Schuster, 1970), 320.
16. Stephen Farber, "The Writer in American Films." *Film Quarterly* 21, no. 4 (Summer 1968): 4.
17. Ibid., 9.
18. In 1979, *Petulia* was listed as the third most significant film of the prior decade—behind *The Godfather* films, jointly named as #1, and *Nashville* at #2—despite the fact that the reviewers stretched the rules by a year to include the film. Both Andrew Sarris and Stephen Farber included *Petulia* as the first choice on their lists. James Monaco, *American Film Now: The People, the Power, the Money, the Movies* (New York: Plume/New American Library, 1979), 449–451.
19. James Price, "Petulia," *Sight & Sound* 37, no. 3 (Summer 1968): 154.
20. Stephen Farber, "Petulia," *Film Quarterly* 22, no. 1 (Fall 1968): 68.
21. What I refer to as fantasy flashes can be read as either flashbacks or flashforwards in the context of the filmic experience, but which are revealed to be fantasies by the end of the story.

22. Gelmis, *Superstar*, 240–241.
23. Stanley Kauffmann, "Lesser But Lester," *New Republic* 158, no. 26 (29 June 1968): 33.
24. Ibid.
25. Phillips, "John Schlesinger: Social Realist," 61.
26. Sound theorist Michel Chion defines synchresis as "the spontaneous and irresistible weld produced between a particular auditory phenomenon and visual phenomenon when they occur at the same time." The simultaneity of the two events lets the audience hear the sound as being produced by the visual object "independent of any rational logic." Synchresis allows sound editors to use myriad different sounds in lieu of the actual sound produced by an object in order to create an aesthetic or emotional effect. Michel Chion, *Audio-Vision: Sound on Screen*, ed. and trans. Claudia Gorbman (New York: Columbia University Press, 1994), 63.
27. Gene Phillips, "John Schlesinger interview," *Film Comment* 11, no. 3 (May–June 1975): 42.
28. Ibid.

CHAPTER 3 TV AND DOCUMENTARY'S INFLUENCE ON SOUND AESTHETICS

1. Some directors who started earlier in the 1960s, like Mike Nichols and Sam Peckinpah, still preferred to rely on the flexibility that looping provided for manipulating the story in post-production. In general, however, there was a major trend away from looping in favor of using location dialogue recordings.
2. "Academy Award-Winning Nagra Recorder," *American Cinematographer* 47, no. 6 (June 1966): 409–411, 432.
3. See Ian Cameron and Mark Shivas, "Cinéma Vérité: New Methods, New Approach," *Movie* no. 8 (April 1963): 12–15.
4. Ray Carney, ed., *Cassavetes on Cassavetes* (New York: Faber and Faber, 2001), 180–181.
5. Al Ruban, "*Faces* from My Point of View," in John Cassavetes, *Faces*, comp. Al Ruban (New York: New American Library, 1970), 13.
6. Carney, *Cassavetes*, 243.
7. Ibid., 342.
8. *David Holzman's Diary* was shot during the summer of 1967 and had its debut at the Mannheim International Film Festival in Germany later that year, where it won the grand prize. Despite its early *succès d'estime* and other festival screenings, the film was unable to find an American distributor until its New York City debut in December 1973. See Nora Sayre, "'David Holzman's Diary' Spoofs Cinema Verite," *New York Times* (7 December 1973): 34.
9. Peter Hogue, "Images," *Film Comment* 29, no. 6 (November–December 1993): 2.
10. Lester D. Friedman, *Bonnie and Clyde* (London: BFI Publishing, 2000), 10.
11. Ibid., 13.
12. A shrewd decision that would make Beatty a substantial fortune when the film went on to gross over $24 million. Friedman, *Bonnie and Clyde*, 14.
13. Joseph Gelmis, *The Film Director as Superstar* (Garden City, NY: Doubleday, 1970), 221.
14. Ibid., 212.
15. Vincent LoBrutto, *Selected Takes: Film Editors on Editing* (Westport, CT: Praeger, 1991), 75.

16. Ibid.
17. Ibid., 76.
18. Paul Monaco, *The Sixties: 1960–1969* (Berkeley: University of California Press, 2003), 107.
19. LoBrutto, *Selected Takes*, 77.
20. Ibid., 78.
21. Gelmis, *Superstar*, 223.
22. LoBrutto, *Selected Takes*, 81.
23. David Prince et al., "Film Editing: An Interview with Dede Allen," *Wide Angle* 2, no. 1 (1977): 63.
24. Especially Fritz Lang's *You Only Live Once* (1937), Nicholas Ray's *They Live by Night* (1949), Don Siegel's *Baby Face Nelson* (1957), and Roger Corman's *Machine-Gun Kelly* (1958).
25. Allan Arkush, "I Want My KEM-TV," *American Film* 11, no. 3 (December 1985): 66.
26. Robin Wood, *Arthur Penn* (New York: Praeger, 1970), 76.
27. Pauline Kael, "Bonnie and Clyde," in *Kiss Kiss Bang Bang* (Boston: Little, Brown, 1968), 49.
28. Robert Shelton, "Record Companies Are Plugging 'Bonnie and Clyde,'" *New York Times* (11 March 1968): 46.
29. Although the trend for sound track albums having dialogue between songs returned in the 1990s, especially in the dialogue and music-heavy films of Quentin Tarantino, it was a novel concept for 1968. Other albums that followed, such as the sound tracks for *Head*, *M*A*S*H*, and *200 Motels*, also mixed dialogue with music from the film, and the practice had an impact on the way that narration and popular music began to intertwine in the early 1970s.
30. See Ruthe Stein, "The Youth Phenomenon in Films," *Cinéaste* 3, no. 2 (Fall 1969): 13–16, 35.
31. Gulf + Western bought Paramount Pictures in 1966, and the irony of the multinational petrochemical company's ownership of the studio clearly is not lost as Wexler makes an explicit connection between Gulf + Western and the grotesque auto accident in the opening shot of the film.
32. Ernest Callenbach and Albert Johnson, "The Danger Is Seduction: An Interview with Haskell Wexler," *Film Quarterly* 21, no. 3 (Spring 1968): 12.
33. Stanley Kauffmann, "*Medium Cool*," *New Republic* (20 September 1969): 20.
34. Richard Corliss, "Haskell Wexler's Radical Education," *Film Quarterly* 23, no. 2 (Winter 1969–70): 53.
35. Herb A. Lightman, "The Filming of *Medium Cool*," *American Cinematographer* 51, no. 1 (January 1970): 23.
36. David Steigerwald, *The Sixties and the End of Modern America* (New York: St. Martin's Press, 1995), 88.
37. Ibid., 89.
38. Paul Cronin, "Mid-Summer Mavericks," *Sight & Sound*, ser. 2, vol. 11, no. 9 (September 2001): 26.
39. Larry Gross, "Look Back—Haskell Wexler's *Medium Cool*," *Millimeter* 4, no. 9 (September 1976): 73.

40. "Par Plays It 'Cool' with - - - - Words," *Daily Variety* (16 July 1969): 14.
41. "Par's 'Medium Cool' Rated X," *Hollywood Reporter* (2 July 1969): n.p. It should be noted that Jack Valenti previously had been a special advisor to President Johnson before he was asked to head the MPAA in 1966. Later, Wexler was able to corroborate the fact that the FBI had actually assembled an extensive dossier on his activities as a documentary filmmaker, and that some of those files may have also effected the near non-release of the film by Paramount. Dennis Schaefer and Larry Salvato, *Masters of Light: Conversations with Contemporary Cinematographers* (Berkeley: University of California Press, 1984), 260.
42. Jay Cocks, "Dynamite," *Time* 94, no. 8 (22 August 1969): 81.
43. Aside from his work as a fashion photographer, Schatzberg was interested in documenting the popular music scene of the 1960s. In addition to photo sessions with The Beatles, Jimi Hendrix, Frank Zappa, and The Rolling Stones, Schatzberg is best known for his photograph of Bob Dylan used for the cover of *Blonde on Blonde*.
44. Faye Dunaway bears a strong physical similarity to Edie Sedgwick, and Thomas Schur points out that the name "Lou Andreas Sand" is a direct allusion to Lou Andreas-Salomé, the lover of Rainer Maria Rilke and close friend of Friedrich Nietzsche and Sigmund Freud. See Thomas Schur, "Faye Dunaway: Stardom and Ambivalence," in *Hollywood Reborn: Movie Stars of the 1970s*, ed. James Morrison (New Brunswick, NJ: Rutgers University Press, 2010), 138–157.
45. There is a direct link between the European New Waves and *Puzzle of a Downfall Child* through the figure of Carole Eastman. She was the original screenwriter for *Petulia* before her script was rejected by producer Raymond Wagner and the job was given to Lawrence B. Marcus. Eastman also worked with *nouvelle vague* director Jacques Demy, providing the English dialogue for his only American film, *The Model Shop* (1969), before her breakthrough work with Bob Rafelson on *Five Easy Pieces* (1970).
46. Especially *Klute*, which appeared the following year and also incorporated the trope of the tape recorder serving as an apparatus for the control and objectification of women. See chapter 4.
47. Culminating, most famously, with the Watergate bugging of the Democratic National Committee headquarters and the secret White House tapes recorded in the Oval Office.
48. This technique is similar to the one-sided conversation previously mentioned in *Point Blank*.
49. Technically this is not an edit. Schatzberg placed a photograph in front of the camera and quickly removed it to achieve the startling effect.
50. Walla is the term used for background ambiences constructed out of human voices, such as the murmur of crowds.
51. Schur, "Faye Dunaway: Stardom and Ambivalence," 142–143.

CHAPTER 4 NEW VOICES AND PERSONAL SOUND AESTHETICS, 1970–1971

1. Aljean Harmetz, "The 15th Man Who Was Asked to Direct *M*A*S*H* (and Did) Makes a Peculiar Western," in *Robert Altman Interviews*, ed. David Sterritt (Jackson: University of Mississippi Press, 2000), 4.
2. Robert Phillip Kolker, *A Cinema of Loneliness: Penn, Kubrick, Scorsese, Spielberg, Altman*, 2nd ed. (New York: Oxford University Press, 1988), 313; emphsis in original.

3. See André Bazin, "Construction in Depth (1950)," in *Orson Welles: A Critical View*, trans. Jonathan Rosenbaum (New York: Harper and Row, 1978), 75–80, and "William Wyler, or the Jansenist of Directing (1951)," *New Orleans Review* 12, no. 4 (Winter 1985); reprinted in *Bazin at Work: Major Essays and Reviews from the Forties and Fifties*, ed. Bert Cardullo, trans. Alain Piette and Bert Cardullo (New York: Routledge, 1997), 1–22.
4. Harmetz, "The 15th Man," 7.
5. Several members of both groups were previously brought in to provide improvisatory elements in Richard Lester's *Petulia* two years prior and many became regulars in Altman's extended stable of actors.
6. David Denby, *Film 70/71: An Anthology by the National Society of Film Critics*, ed. David Denby (New York: Simon and Schuster, 1971), 78.
7. Mike Steen, *Hollywood Speaks: An Oral History* (New York: Putnam, 1974), 323.
8. Harmetz, "The 15th Man," 8.
9. Robin Wood, *Hollywood from Vietnam to Reagan* (New York: Columbia University Press, 1986), 35.
10. John Belton, "The Bionic Eye: Zoom Esthetics," *Cineaste* 11, no. 1 (Winter 1980–81): 25.
11. Steen, *Hollywood Speaks*, 325.
12. Jim Webb, telephone interview with the author, 12 February 2001.
13. *M*A*S*H* was one of the first films to feature the graphic effect of blood splashing onto the camera lens.
14. Kolker, *Cinema of Loneliness*, 2nd ed., 312.
15. See Robert T. Self, "The Sounds of *M*A*S*H*," in *Close Viewings: An Anthology of New Film Criticism*, ed. Peter Lehman (Gainesville: University Press of Florida, 1990), 141–157.
16. Brother Alexis, "Musing on *M*A*S*H*—An Interview with Robert Altman," *New Orleans Review* 2, no. 2 (1970): 119.
17. Belton, "The Bionic Eye," 29.
18. Michel Chion, *Audio-Vision: Sound on Screen*, ed. and trans. Claudia Gorbman (New York: Columbia University Press, 1994), 72.
19. Fredric Jameson, *The Geopolitical Aesthetic* (London: BFI Publishing, 1992), 52–56.
20. Ibid., 20.
21. Michel Chion, *The Voice in Cinema*, trans. Claudia Gorbman (New York: Columbia University Press, 1999), 17–29.
22. Jameson, *Geopolitical Aesthetic*, 43; emphasis in original.
23. Chion, *Voice in Cinema*, 126.
24. This is an interesting intersection between life and art as Pakula studied to be a therapist before entering into the film industry. In order to get the dialogue for these scenes, Pakula hired a real psychologist to play the part of the therapist and to "interview" Jane Fonda in character. The result was an improvised dialogue between the two actors as well as a detailed analysis behind the motivation of Bree's character. Tom Milne, "Not a Garbo or a Gilbert in the Bunch: Alan Pakula talks to Tom Milne," *Sight & Sound* 41, no. 2 (Spring 1972): 88–93.
25. Robin Wood, "*Klute*," *Film Comment* 8, no. 2 (Spring 1972): 34.
26. Diane Giddis, "The Divided Woman: Bree Daniels in *Klute*," *Women and Film* 1, nos. 3 and 4 (1973): 59.

27. Milne, "Not a Garbo," 92.
28. Stephen Paul Miller, *The Seventies Now: Culture as Surveillance* (Durham, NC: Duke University Press, 1999), 124.
29. Thomas Elsaesser, "Notes on the Unmotivated Hero—The Pathos of Failure: American Films in the 70's," *Monogram* no. 6 (October 1975): 19; my emphesis.
30. See Ruthe Stein, "The Youth Phenomenon in Films," *Cinéaste* 3, no. 2 (Fall 1969): 13–16, 35.
31. Lee Hill, *Easy Rider* (London: BFI, 1996), 26.
32. For more on the compilation score, see chapter 8.
33. Mitchell S. Cohen, "The Corporate Style of BBS," *Take One* 3, no. 12 (July–August 1972): 19–22.
34. Shelley Benoit, "Prototype for Hollywood's New Freedom," *Show* 2, no. 1 (March 1971): 24–26.
35. David A. Cook, *Lost Illusions: American Cinema in the Shadow of Watergate and Vietnam, 1970–1979* (Berkeley: University of California Press, 2000), 312. For more on *American Graffiti*, see chapter 8.
36. Shelley Benoit, "On the Road with the New Hollywood," *Show* 2, no. 1 (March 1971): 19.
37. In October 1970, Taylor's single "Fire and Rain," from his second album *Sweet Baby James* (released in February), cracked the Top Ten of the pop charts and made him one of the most sought-after performers even though he was on the road shooting *Two-Lane Blacktop*. By the time of the film's debut in July 1971, Taylor already had another hit single with "Country Road," appeared on the cover of *Time* magazine, and released a third album, *Mud Slide Slim*.
38. Benoit, "On the Road," 17.
39. Hellman chose to use only a few seasoned actors in the film, primarily in small roles. Nearly all the other actors and extras were people they recruited as the production traveled from location to location across the Southwest and South. Oates's character name refers to the 1970 Pontiac GTO he drives in the film.
40. The compact cassette tape, introduced by Philco in the late 1960s, did not achieve a commercial status until the mid-1970s. In 1970, the time when the film was shot, over 90 percent of the installations of automobile tape decks used the Lear eight-track cartridge. Because the GTO in the film is outfitted with a built-in cassette deck, it emphasizes the "newness" of the car and its owner's vain effort to have every conceivable accessory. Russell Sanjek, *Pennies from Heaven: The American Popular Music Business in the Twentieth Century* (New York: Da Capo, 1996), 550.
41. Adam Webb, "Two Lane Blacktop," in *Lost Highways: An Illustrated History of Road Movies*, ed. Jack Sargent and Stephanie Watson (London: Creation Books, 1999), 83.
42. The fact that the music licensing department did not see fit to arrange for music rights other than the film and television release also caused the extended delay of *Two-Lane Blacktop*'s video release. See Melissa Pierson, "Starting Over, and Over: The Curious Career of Monte Hellman," *Village Voice* (6 August 1991): 59–60, and Charles Lyons, "Moonlight Drive: 'Two Lane' on Road to DVD, Vidbin at Last," *Variety* (27 September 1999): 7.
43. Christian Braad Thomsen, "Monte Hellman: A Profile," *Take One* 4, no. 2 (November–December 1972): 30.

44. Aljean Harmetz, "Monte's Turn for the Big Time?" *New York Times* (16 May 1971): 11.
45. Gary Kurtz, "Commentary," *Two-Lane Blacktop*, directed by Monte Hellman (1973; Beverly Hills: Anchor Bay Entertainment, 1999), DVD.
46. Ibid.
47. Monte Hellman, "Commentary," *Two-Lane Blacktop*, DVD.
48. Thomsen, "Monte Hellman," 30.
49. For more on *Star Wars* and Dolby Stereo, see chapter 9.
50. Walter Murch, "Dense Clarity—Clear Density" (paper presented at the "Walter Murch and the Art of Sound Design" conference, University of Iowa, Iowa City, 29 March 2000).
51. Entirely coincidentally, in 1970 Dolby Laboratories opened their U.S. operations in San Francisco just down the block from American Zoetrope. There is no doubt that the presence of such a vibrant filmmaking community in the San Francisco area led Dolby Laboratories to investigate the possibility of utilizing Dolby noise reduction to improve the quality of motion picture sound.
52. Christopher Pearce, "San Francisco's Own American Zoetrope," *American Cinematographer* 52, no. 10 (October 1971): 1004.
53. Walter Murch, "The Special Sounds of *THX 1138*," *American Cinematographer* 52, no. 10 (October 1971): 1075.
54. Vincent LoBrutto, *Sound-on-Film: Interviews with Creators of Film Sound* (Westport, CT: Praeger, 1994), 85.
55. Ibid.
56. Walter Murch, telephone interview with the author, 23 January 2001.
57. LoBrutto, *Sound-on-Film*, 87.
58. Walter Murch, "Sound Design: The Dancing Shadow," in *Projections 4: Film-makers on Film-making*, ed. John Boorman, Tom Luddy, David Thomson, and Walter Donohue (London: Faber and Faber, 1995), 245. For more on *The Rain People* and Murch's collaboration with Coppola, see chapter 5.
59. Murch, "Special Sounds," 1075.
60. Michael Pye and Lynda Myles, *The Movie Brats: How the Film Generation Took Over Hollywood* (New York: Holt, Rinehart and Winston, 1979), 116.
61. Cook, *Lost Illusions*, 138.

CHAPTER 5 FRANCIS FORD COPPOLA'S AMERICAN ZOETROPE AND COLLECTIVE FILMMAKING

1. Jeffrey Chown, *Hollywood Auteur: Francis Coppola* (New York: Praeger, 1988), 51.
2. Michael Pye and Lynda Myles, *The Movie Brats: How the Film Generation Took Over Hollywood* (New York: Holt, Rinehart and Winston, 1979), 83.
3. Michael Ondaatje, *The Conversations: Walter Murch and the Art of Editing Film* (New York: Alfred A. Knopf, 2002), 20.
4. Diane Jacobs, "Look Back: Francis Ford Coppola's *The Rain People*," *Millimeter* 4, no. 11 (November 1976): 61.
5. See Michel Chion, *Audio-Vision: Sound on Screen*, ed. and trans. Claudia Gorbman (New York: Columbia University Press, 1994), 87–88.

6. Peter Cowie, *Coppola* (New York: Charles Scribner's Sons, 1990), 47.
7. Fabrice Moussus, "The Art of the Boom: Nat Boxer," *Filmmakers Newsletter* 9, no. 7 (May 1976): 19.
8. Ibid., 20.
9. Ibid., 18.
10. Ibid., 20.
11. Cowie, *Coppola*, 48.
12. Chion, *Audio-Vision*, 25–28.
13. Ibid., 29.
14. Michel Chion, *Film, A Sound Art*, trans. Claudia Gorbman (New York: Columbia University Press, 2009), 142.
15. Robert Phillip Kolker, *A Cinema of Loneliness: Penn, Kubrick, Coppola, Scorsese, Altman* (New York: Oxford University Press, 1980), 169.
16. Ondaatje, *The Conversations*, 125; emphasis in original.
17. Rick Altman, "Introduction: Four and a Half Film Fallacies," in *Sound Theory/Sound Practice*, ed. Rick Altman (New York: AFI/Routledge, 1992), 35–45.
18. Ondaatje, *The Conversations*, 121.
19. Ibid.
20. Judith Vogelsang, "Motifs of Image and Sound in *The Godfather*," *Journal of Popular Film* 2, no. 2 (Spring 1973): 125; emphasis in original.
21. Cowie, *Coppola*, 74.
22. Ondaatje, *The Conversations*, 102–103.
23. Ibid., 98.
24. Chion, *Film*, 142.
25. Cowie, *Coppola*, 96.
26. Kolker, *Cinema of Loneliness*, 198.
27. Ibid.
28. Walter Murch, "Foreword" [1994], in Chion, *Audio-Vision*, xxii; my emphasis.
29. David Denby, "Stolen Privacy: Coppola's *The Conversation*," *Sight & Sound* 43, no. 3 (Summer 1974): 131–133.

CHAPTER 6 ROBERT ALTMAN'S COLLABORATIVE SOUND WORK

1. Bruce Williamson, "*Playboy* Interview: Robert Altman," *Playboy* (August 1976): 61.
2. Robert Phillip Kolker, *A Cinema of Loneliness: Penn, Kubrick, Scorsese, Spielberg, Altman*, 2nd ed. (New York: Oxford University Press, 1988), 313.
3. C. Kirk McClelland, *Brewster McCloud: A Day-by-Day Journal of the On-Location Shooting of the Film in Houston* (New York: New American Library, 1971), 22.
4. Mitchell Zuckoff, *Robert Altman: The Oral Biography* (New York: Vintage Books, 2010), 219.
5. Ibid.
6. John Cutts, "*M*A*S*H*, *McCloud*, and *McCabe*—An Interview with Robert Altman," *Films and Filming* 18, no. 2 (November 1971): 44.
7. Stuart Rosenthal, "McCabe and Mrs. Miller," *Focus on Film* 9 (Spring 1972): 9.

8. Paul Arthur, "How the West Was Spun: *McCabe & Mrs. Miller* and Genre Revisionism," *Cineaste* 28, no. 3 (Summer 2003): 19.
9. David Thompson, ed., *Altman on Altman* (New York: Faber and Faber, 2005), 73.
10. Judith M. Kass, *Robert Altman: American Innovator* (New York: Popular Library, 1978), 126.
11. Donald Lyons, "Flaws in the Iris: The Private Eye in the Seventies," *Film Comment* 29, no. 4 (July–August 1993): 49.
12. Kass, *Robert Altman*, 156.
13. Zuckoff, *Robert Altman*, 250.
14. Ibid.
15. Thompson, *Altman on Altman*, 80.
16. Ibid.
17. Kass, *Robert Altman*, 158.
18. Thompson, *Altman on Altman*, 83.
19. Kolker, *Cinema of Loneliness*, 2nd ed., 336.
20. James E. Webb, "Multi-Channel Dialogue and Effects Recording During Film Production," *American Cinematographer* 60, no. 4 (April 1979): 368.
21. Jim Webb, telephone interview with the author, 12 February 2001.
22. Bruce Berman, "California Compulsions," *1000 Eyes Magazine* 6 (January 1976): 18.
23. Jim Webb, with assistance from Don Ketteler, "Using the Multi-Track Format for Production Film Recording," *Recording Engineer/Producer* 11, no. 3 (1980): 117.
24. See Rick Altman, "Sound Space," in *Sound Theory/Sound Practice* (New York: AFI/Routledge, 1992), 46–64.
25. Kass, *Robert Altman*, 189.
26. Kolker, *Cinema of Loneliness*, 2nd ed., 353.
27. For more on Altman's sound techniques in his early films, see my chapter "The Democratic Voice—Robert Altman's Sound Aesthetics in the 1970s" in *A Companion to Robert Altman*, ed. Adrian Danks (Hoboken, NJ: Wiley-Blackwell, 2015), 184–209.
28. Sid Levin, "The Art of the Editor: *Nashville*," *Filmmakers Newsletter* 8, no. 10 (August 1975): 32.
29. Thomas Elsaesser, "Notes on the Unmotivated Hero—The Pathos of Failure: American Films in the 70's," *Monogram* no. 6 (October 1975): 14.
30. Voiced by Thomas Hal Phillips, whose Tennessee accent sounded curiously close to that of Howard Baker, then serving his second term in the Senate and familiar to many Americans from his participation in the Senate committee investigating the Watergate scandal.
31. Rick Altman, "24-Track Narrative? Robert Altman's *Nashville*," *Cinéma(s)* 1, no. 3 (1991): 103.
32. Ibid., 120.
33. Kolker, *Cinema of Loneliness*, 2nd ed., 351.

CHAPTER 7 MARTIN SCORSESE'S DIALECTICAL SOUND

1. This does not mean that Scorsese's sound tracks were inconsistent in their aesthetics, but just the opposite. Due to the vicissitudes of working with several different studios

(AIP, Warner Bros., Columbia, United Artists) and shooting in New York as well as Los Angeles, it proved necessary to rely on different sound recordists, editors, and mixers for each picture. To ensure a consistency from film to film and to utilize the sound track to its fullest potential, Scorsese was present in all phases of post-production sound and would personally supervise all looping sessions and select the music for each of his films. This is documented in several files of handwritten notes and video recordings of the director's work, available in the Martin Scorsese Collection at the American Film Institute's Louis B. Mayer Library.

2. James Powers, "Martin Scorsese Seminar," *Dialogue on Film* 4, no. 7 (April 1975): 9.
3. For more on the use of music in *American Graffiti*, see chapter 8.
4. At the time when Anger was making his film nearly all the songs on the sound track would have been considered rock 'n' roll. However, in retrospect, the split between the white and black strains of popular music in the 1960s can start to be heard in the R&B of "Hit the Road Jack" and the Motown sound of Martha and the Vandellas' "Heat Wave."
5. Ed Lowry, "The Appropriation of Signs in *Scorpio Rising*," *Velvet Light Trap* no. 20 (Summer 1983): 42.
6. Scorsese recounted his first viewing of *Scorpio Rising*, when Jonas Mekas screened the film in 1964: "Of all his films, it is still the one I like the most. I was impressed with his ironic use of pop music and by his sense of rhythm. I was also fascinated by his ambiguous relationship to Hollywood, his love/hate continuously expressed through the use and abuse of Hollywood imagery, in a romantic, decadent mood." Quoted in Jayne Pilling and Mike O'Pray, eds., *Into the Pleasure Dome: The Films of Kenneth Anger* (London: BFI, 1989), 55.
7. Robert Phillip Kolker, *A Cinema of Loneliness: Penn, Kubrick, Scorsese, Spielberg, Altman*, 2nd ed. (New York: Oxford University Press, 1988), 166.
8. Ian Penman, "Juke Box and Johnny Boy," *Sight & Sound* ser. 2, vol. 3, no. 4 (April 1993): 10.
9. Ibid.
10. See Claudia Gorbman, *Unheard Melodies: Narrative Film Music* (Bloomington: Indiana University Press, 1987), 22–23.
11. Pauline Kael, "Everyday Inferno," *New Yorker* (8 October 1973): 162.
12. Martin Scorsese, "Mean Streets Music," handwritten note, c. 1973, 1–2, in Scorsese collection, American Film Institute.
13. Ibid.; Lois McGrew, "Memo to Barry Haldeman, Rosenfeld, Meyer & Susman—Re: Mean Streets," 9 April 1973, 1, in Scorsese collection, American Film Institute.
14. The fact that the song became a hit after Ace had died from a self-inflicted gunshot wound while playing Russian roulette may have also been intended by Scorsese to create a sense of foreboding in the scene.
15. Scorsese's handwritten notes indicate that the original choice of song was "Hide Away" by John Mayall and the Bluesbreakers, another band featuring Eric Clapton. As Scorsese mentioned to James Powers, "The music was always in the original script," and the dialectic between rock and doo-wop was at the heart of the film's narrative. Powers, "Martin Scorsese Seminar," 7.
16. Andrew C. Bobrow, "The Filming of *Mean Streets*," *Filmmakers Newsletter* 7, no. 3 (January 1974): 28.

17. Ibid., 29.
18. "ADR Sheets" [*Alice Doesn't Live Here Anymore*], 26 August 1974, 1; Virginia Chase, "Memo to Rudi Fehr—Subject: ALICE DOESN'T LIVE HERE ANYMORE—Revised Looping Schedule," 26 September 1974, 1, in Scorsese collection, American Film Institute.
19. F. Anthony Macklin, "'It's a Personal Thing for Me': An Interview with Martin Scorsese," *Film Heritage* 10, no. 3 (Spring 1975): 18–19; Paul Gardner, "Martin Scorsese," *Action!* 10, no. 3 (May–June 1975): 34.
20. Michael Pye and Lynda Myles, *The Movie Brats: How the Film Generation Took Over Hollywood* (New York: Holt, Rinehart and Winston, 1979), 204.
21. Karyn Kay and Gerald Peary, "*Alice Doesn't Live Here Anymore*: Waitressing for Warner's," *Jump Cut* no. 7 (May–July 1975): 6.
22. James Monaco, *American Film Now: The People, the Power, the Money, the Movies* (New York: Plume/New American Library, 1979), 156.
23. Steve Howard, "The Making of *Alice Doesn't Live Here Anymore*: An Interview with Director Martin Scorsese," *Filmmakers Newsletter* 8, no. 5 (March 1975): 25.
24. Ibid.
25. Ibid.
26. Russell E. Davis, "*Alice Doesn't Live Here Anymore*: Under the Comic Frosting," *Jump Cut* no. 7 (May–July 1975): 3.
27. Brian Henderson, "*The Searchers*: An American Dilemma," *Film Quarterly* 34, no. 2 (Winter 1980–81): 9–10; Noël Carroll, "The Future of Allusion: Hollywood in the Seventies (and Beyond)," *October* no. 20 (Spring 1982): 65–66.
28. Michael Dempsey, "Taxi Driver," *Film Quarterly* 29, no. 4 (Summer 1976): 37.
29. "Suspension occurs when a sound naturally expected from a situation (which we usually hear at first) becomes suppressed, either insidiously or suddenly. This creates an impression of emptiness or mystery, most often without the spectator knowing it; the spectator feels its effect but does not consciously pinpoint its origin." Michel Chion, *Audio-Vision: Sound On Screen*, ed. and trans. Claudia Gorbman (New York: Columbia University Press, 1994), 132.
30. Ann Powell, "Scorsese and His Saint," *Millimeter* 4, no. 3 (March 1976): 31.
31. Ibid., 31–32.
32. Ibid., 32.
33. Leighton Grist, *The Films of Martin Scorsese, 1963–77* (New York: St. Martin's Press, 2000), 136–137.
34. Patricia Patterson and Manny Farber, "The Power and the Gory," *Film Comment* 12, no. 3 (May-June 1976): 30.
35. Powell, "Scorsese and His Saint," 30.
36. Ibid., 32.
37. Grist, *Films of Martin Scorsese, 1963–77*, 139.
38. Ibid., 141.
39. Patterson and Farber, "The Power and the Gory," 27.
40. Charles Michener, "Taxi Driver," *Film Comment* 12, no. 2 (March–April 1976): 5; emphasis in original.

41. Kolker, *Cinema of Loneliness*, 2nd ed., 185.
42. Vincent LoBrutto, *Sound-on-Film: Interviews with Creators of Film Sound* (Westport, CT: Praeger, 1994), 34.
43. See Chion, *Audio-Vision*, 86–89.
44. LoBrutto, *Sound-on-Film*, 34.
45. Ibid., 33.
46. Ibid., 34.
47. Grist, *Films of Martin Scorsese, 1963–77*, 151.

CHAPTER 8 THE SOUND OF MUSIC

1. James Monaco, *American Film Now: The People, the Power, the Money, the Movies* (New York: Plume/New American Library, 1979), 133–134.
2. See Stephen Farber, "Movies from Behind the Barricades," *Film Quarterly* 24, no. 2 (Winter 1970/71): 24–33; Ruthe Stein, "The Youth Phenomenon in Films," *Cinéaste* 3, no. 2 (Fall 1969): 13–16, 35.
3. Michael Pye and Lynda Myles, *The Movie Brats: How the Film Generation Took Over Hollywood* (New York: Holt, Rinehart and Winston, 1979), 121.
4. Ibid.
5. Larry Sturhahn, "The Art of the Sound Editor: An Interview with Walter Murch," *Filmmakers Newsletter* 8, no. 2 (December 1974): 24.
6. Jeff Smith, *The Sounds of Commerce: Marketing Popular Film Music* (New York: Columbia University Press, 1998), 174.
7. Vincent LoBrutto, *Sound-on-Film: Interviews with Creators of Film Sound* (Westport, CT: Praeger, 1994), 88.
8. See "Sensurround," *American Cinematographer* 55, no. 10 (November 1974): 1312–1313, 1345–1353; Joseph McBride, "Universal Exploring Possibility of Expanding Sensurround Use," *Variety* (12 December 1974): 10.
9. Pye and Myles, *The Movie Brats*, 127.
10. Ibid.; my emphasis.
11. Irene Kahn Atkins, "The Melody Lingers On: Source Music in Films of the American Past," *Focus on Film* no. 26 (1977): 31.
12. See Smith, *Sounds of Commerce*, 172–185; David R. Shumway, "Rock 'n' Roll Sound Tracks and the Production of Nostalgia," *Cinema Journal* 38, no. 2 (Winter 1999): 39–43.
13. Smith, *Sounds of Commerce*, 173.
14. Monaco, *American Film Now*, 169.
15. See Thomas Weiner, "The Rise and Fall of the Rock Film: From *Woodstock* to *Stardust*, the Parade's Gone By," *American Film* 1, no. 3 (December 1975): 58–63; Ivor Davis, "Golddiggers of 1978," *Los Angeles Magazine* (December 1977): 215–217, 280–284; Janet Maslin, "Rock Musicals: An End of an Art Form," *New York Times* (17 June 1978): 32; Dave Marsh, "Schlock around the Rock," *Film Comment* 14, no. 4 (July–August 1978): 7–13; and William P. Kelly, "Running on Empty: Reimagining Rock and Roll," *Journal of American Culture* 4, no. 4 (Winter 1981): 152–159.

16. Joseph McBride, "Song for Woody," *Film Comment* 12, no. 6 (November–December 1976): 27.
17. Choosing to start the film with a song from The Beach Boys creates a curious counterpoint to *American Graffiti*, which concludes with their 1964 hit "All Summer Long." "Wouldn't It Be Nice" provides a pivot point from the halcyon days of the early 1960s to the decline of the counterculture and election of Richard Nixon in 1968 as represented in *Shampoo*.
18. David Ehrenstein, "Shampoo," *Film Quarterly* 28, no. 4 (Summer 1975): 64.
19. William Avery, "Hal Ashby—Getting the Right Drift," *Soho Weekly News* (23 February 1978): 19.
20. Peter Occiogrosso, "Reelin' and Rockin'," *American Film* 9, no. 6 (1 April 1984): 46.
21. Ibid., 47–48.
22. Ibid., 48.
23. See Albert and David Maysles, "*Gimme Shelter*: Production Notes," *Filmmakers Newsletter* 5, no. 2 (December 1971): 30–31.
24. See Jay Beck, "A Quiet Revolution: Changes in American Film Sound Practices, 1967–1979" (PhD diss., University of Iowa, 2003), 147–156.
25. See John Belton, "1950s Magnetic Sound: The Frozen Revolution," in *Sound Theory/ Sound Practice*, edited by Rick Altman (New York: Routledge, 1992), 154–167.
26. See chapter 2 in Beck, "A Quiet Revolution."
27. Ioan Allen, "The Dolby Sound System for Recording *Star Wars*," *American Cinematographer* 58, no. 7 (July 1977): 709.
28. Jay Beck, "Dolby Laboratories," in *The Encyclopedia of Stanley Kubrick*, ed. Gene D. Phillips and Rodney Hill (New York: Checkmark Books, 2002), 79–81.
29. Larry Blake, *Film Sound Today: An Anthology from Recording Engineer/Producer* (Hollywood: Reveille Press, 1984), 9.
30. *A History of Dolby Laboratories* (San Francisco: Dolby Laboratories, 2000), 3.
31. See Ronald E. Uhlig, "Stereophonic Photographic Soundtracks," *Journal of the SMPTE* 82, no. 4 (April 1973): 292–295.
32. See "Dolby Encoded High-Fidelity Stereo Optical Sound Tracks," *American Cinematographer* 56, no. 9 (September 1975): 1032–1033, 1088–1090; Ronald E. Uhlig, "The Sound of the Story," *American Cinematographer* 59, no. 8 (August 1978): 782–783, 786, 813–815.
33. The matrix processor was a way of using phase-change relationships between the two audio channels on the optical sound track to encode a further two channels of information—a center channel and a surround channel.
34. *The Evolution of Dolby Film Sound* (San Francisco: Dolby Laboratories, 2000), 2–3.
35. Even though Dolby Stereo was applied to the optical prints of *Tommy* for its London debut in February 1975, the film was recorded and mixed with the intention of using John Mosely's "Quintaphonic" sound system. See Beck, "A Quiet Revolution," 146–156. Although there is no direct evidence to support the claim, it can be speculated that Mosely's system was not fully operational at the time of the London debut. However, Mosely's "Quintaphonics" was featured in all American print advertisements and reviews the following month. Dolby Stereo did not have its official American debut until November 1975 with the release of *Lisztomania*.

36. *A Chronology of Dolby Laboratories: May 1965–May 1999* (San Francisco: Dolby Laboratories, 1999), 3.
37. Ioan Allen, telephone interview with the author, 23 April 2001.
38. Ibid.
39. Allen, "Dolby Sound System for Recording *Star Wars*," 748.
40. Blake, *Film Sound Today*, 3–4.
41. *Dolby Surround Mixing Manual* (San Francisco: Dolby Laboratories, 1998), 5–1.
42. Ibid., 5–13.
43. See Bruce P. Bogert, "Stereophonic Sound Reproduction Enhancement Utilizing the Haas Effect," *Journal of the Society of Motion Picture and Television Engineers* 64 (June 1955): 308–309.
44. James Corcoran and Douglas Williams, "The Recording and Re-recording of Stereophonic Sound for Wide-Screen Motion Pictures," *Journal of the SMPTE* 77, no. 12 (December 1968): 1292.
45. Ibid., 1293.
46. Richard Portman, telephone interview with the author, 25 June 2000.
47. Even though the original line-up of the group did not continue to tour, a re-formed version of The Band without Robbie Robertson took to the road again in 1983 until the death of keyboard player Richard Manuel in 1986.
48. Monaco, *American Film Now*, 160–161.
49. J. P. Telotte, "Scorsese's *The Last Waltz* and the Concert Genre," *Film Criticism* 4, no. 2 (Winter 1979): 10.
50. Terrence Rafferty, "Martin Scorsese's Still Life," *Sight & Sound* 52, no. 3 (Summer 1983): 190.
51. Leighton Grist, *The Films of Martin Scorsese, 1978–99: Authorship and Context II* (New York: Palgrave Macmillan, 2013), 37.
52. Telotte, "Scorsese's *The Last Waltz*," 12.
53. David Bartholomew, "The Last Waltz," *Film Quarterly* 33, no. 2 (Winter 1979–1980): 56.
54. Initially created as a concept album by composer Andrew Lloyd Webber and lyricist Tim Rice, the recording was turned into a Broadway musical in 1971. R. Serge Denisoff and William D. Romanowski, *Risky Business: Rock in Film* (New Brunswick, NJ: Transaction Publishers, 1991), 208.
55. Ibid., 213.
56. Eric Breitbart, "Lost in the Hustle: An Interview with John Badham," *Cinéaste* 9, no. 2 (Winter 1978–79): 2–3.
57. Susan Schenker, "Saturday Night Fever," *Take One* 6, no. 4 (March 1978): 11.
58. Denisoff and Romanowski, *Risky Business*, 222.
59. Ibid., 225.
60. No doubt to capitalize on the return of the youth-oriented *Star Wars* to the top spot at the box office in September 1977.
61. Denisoff and Romanowski, *Risky Business*, 227.
62. Ibid., 225, 232–233.
63. James Scale, "Sound Recording: No More Ugly Step-Sister—California," *Millimeter* 7, no. 3 (March 1979): 133.

64. Blake, *Film Sound Today*, 14.
65. Occiogrosso, "Reelin' and Rockin'," 47.

CHAPTER 9 THE SOUND OF SPECTACLE

1. Gerald R. Barrett, "William Friedkin Interview," *Literature/Film Quarterly* 3, no. 4 (Fall 1975): 343.
2. The Oscar was awarded to production mixer Christopher Newman and re-recording mixer Robert "Buzz" Knudson. The irony in this traditional breakdown of the award between production and post-production sound is that it effaced the contribution of several other individuals on the sound team. By means of contrast, the film's nomination for Best Sound Track at the 1975 BAFTA (British Academy of Film and Television Arts) Awards recognized the contributions of Christopher Newman, Jean-Louis Ducarme, Robert Knudson, Fred J. Brown, Bob Fine, Ross Taylor, Ron Nagel, Doc Siegel, Gonzalo Gavira, and Hal Landaker.
3. Ruth McCormick, "'The Devil Made Me Do It!'—A Critique of *The Exorcist*," *Cinéaste* 6, no. 3 (1974): 18.
4. James Monaco, *American Film Now: The People, the Power, the Money, the Movies* (New York: Plume/New American Library, 1979), 148–149.
5. The film was remixed for Dolby Stereo upon its 70mm re-release in 1979 and remastered and remixed in Dolby Digital 5.1-channel for the 1998 twenty-fifth anniversary DVD. More recently, the film has been extended and partially re-recorded for its 2000 theatrical re-release, and remastered for the fortieth anniversary BluRay release. All citations regarding the film's sound refer to the original 1973 monophonic version.
6. Cindy Ehrlich, "Doin' the Devil's Sound Effects," *Rolling Stone* no. 155 (28 February 1974): 16.
7. The use of a magnetic release print may have been suggested to Friedkin by dubbing mixer Buzz Knudson. See William Friedkin, personal note on back of "ADR—*The Exorcist*: Reel no. four, sheet no. one A," 3 October 1973, in the William Friedkin collection, Margaret Herrick Library, Academy of Motion Picture Arts and Sciences, Beverly Hills.
8. Vincent LoBrutto, *Sound-on-Film: Interviews with Creators of Film Sound* (Westport, CT: Praeger, 1994), 199; my emphasis.
9. James Powers, "William Friedkin Seminar," *Dialogue on Film* 3, no. 4 (February/March 1974): 4.
10. Twentieth Century–Fox's widescreen system, CinemaScope, utilized four-track release prints with magnetic striping running to the left and right of the sprocket holes on each side of the filmstrip, a format fully commensurate with the newly adopted three-track dubbers. Because of the space needed for the magnetic stripes, early CinemaScope prints eliminated the optical sound track and could be played only on specially modified projectors with a four-channel magnetic playback head. To ensure that theaters would modify their projectors to accommodate both the new CinemaScope lenses and the four-channel magnetic sound track, in 1953 Fox President Spyros Skouras announced that all Fox CinemaScope prints would be released exclusively in the stereophonic format. "New Sound System Developed by Fox," *New York Times* (12 May 1953): 31.

11. For more on Schifrin's unused score, see Claire Sisco King, "Ramblin' Men and Piano Men: Crises of Music and Masculinity in *The Exorcist*," in *Music in the Horror Film: Listening to Fear*, ed. Neil Lerner (New York: Routledge, 2009), 115–117.
12. Powers, "William Friedkin Seminar," 21.
13. See Michael Hannon, "Sound and Music in Hammer's Vampire Films," in *Terror Tracks: Music, Sound and Horror Cinema*, ed. Phillip Hayward (London: Equinox, 2009), 60–74; and Janet K. Halfyard, "Mischief Afoot: Supernatural Horror-Comedies and the *Diabolus in Musica*," in Lerner, *Music in the Horror Film*, 21–37.
14. William Friedkin, "Exorcist Cues," c. November 1973, in the William Friedkin collection, Margaret Herrick Library, Academy of Motion Picture Arts and Sciences.
15. These cues include "Chris drives home from doctor's office," "Chris & Karl return Regan from the clinic," "Karras walks home from Chris's," "Karras rushes to Chris's house," "Merrin & Karras prepare for the exorcism," and "Karras & Merrin return to Regan's room." Ibid.
16. Ehrlich, "Doin' the Devil's Sound Effects," 16.
17. Ibid.
18. *The Fear of God: The Making of the Exorcist*, BBC-TV, 1998.
19. LoBrutto, *Sound-on-Film*, 63.
20. Chicago-based radio host Ken Nordine, who was not credited in the film, was hired by Friedkin in 1973 to develop a number of the effects for *The Exorcist*. Although the director claimed that none of Nordine's sound effects were used in the film, Nordine filed suit against Warner Bros. on 23 January 1974 to recover his contracted payment of $35,633 and to seek proper screen credit. Although Nordine subsequently did not receive screen credit, he did receive a cash settlement from Warner Bros. in 1979 after it was demonstrated that some of Nordine's sound effects were used in the final film, specifically a number of the animal squeals and "the sound of hamster feet scratching inside a cardboard box." See "Nordine vs. 'Exorcist,' & Vice Versa—Mercedes McCambridge's Demon Voice Part of Warners Counter-Suit over 'Effects,'" *Variety* 294 (7 February 1979): 6, and "WB Settles on Trial's 4th Day," *Variety* 294 (7 February 1979): 6.
21. Bud Smith was hired to edit the Iraq prologue only. This was done so that Smith could edit and supervise the sound construction of Reel 1 while Evan Lottman and Friedkin were editing the rest of the film. See Gabriella Oldham, *First Cut: Conversations with Film Editors* (Berkeley: University of California Press, 1992), 227.
22. Ehrlich, "Doin' the Devil's Sound Effects," 16.
23. Ibid.
24. "Re-recording log sheets, Todd-AO," 15 December 1978, in the William Friedkin collection, Margaret Herrick Library, Academy of Motion Picture Arts and Sciences.
25. *Warner Bros. Inc. and Hoya Productions, Inc. v. Film Ventures International*, United States District Court, Central District of California, No. CV 75–2774-DWW, Civil (5 September 1975), 12, lines 5–10.
26. See chapter 10.
27. See Michel Chion, *Audio-Vision: Sound on Screen*, ed. and trans. Claudia Gorbman (New York: Columbia University Press, 1994), 63–64.
28. *The Fear of God*, BBC-TV.
29. Powers, "William Friedkin Seminar," 9.

30. Michel Chion, *The Voice in Cinema*, trans. Claudia Gorbman (New York: Columbia University Press, 1999), 164, 171.
31. Richard Buskin, "Head-Turning Sound," *Surround Sound Professional* 4, no. 1 (January–February 2001): 33.
32. Charles Higham, "Will the Real Devil Speak Up? Yes!," *New York Times* (27 January 1974): D13.
33. Ibid.
34. One last session to record the "Devil's Voice Wild Tracks" was done on 9 October 1973 with actors Michael Christopher, Phillippa Harris, and Liam Dunn contributing.
35. Powers, "William Friedkin Seminar," 9.
36. William Friedkin, "Letter to the Editor," *New York Times* (28 January 1974): Arts and Leisure section, 2.
37. *Warner Bros. Inc. and Hoya Productions, Inc. v. Film Ventures International*, 11, lines 19–21.
38. Ibid.
39. William Friedkin, "Affidavit—1," filed in support of *Warner Bros. Inc. and Hoya Productions, Inc. v. Film Ventures International* (5 September 1975), 3, lines 17–18, in the William Friedkin collection, Margaret Herrick Library, Academy of Motion Picture Arts and Sciences.
40. *Warner Bros. Inc. and Hoya Productions, Inc. v. Film Ventures International*, 11, lines 29–30; 12, lines 11 and 22.
41. Bud Smith, "Affidavit—3," filed in support of *Warner Bros. Inc. and Hoya Productions, Inc. v. Film Ventures International* (5 September 1975), 22, line 26, in the William Friedkin collection, Margaret Herrick Library, Academy of Motion Picture Arts and Sciences.
42. The case of *Warner Bros. Inc. and Hoya Productions, Inc. v. Film Ventures International* was heard on 10 October 1975. Although Judge David W. Williams ruled that the advertising campaign for *Beyond the Door* improperly suggested that the film was a sequel to *The Exorcist*, he did not find sufficient grounds to rule on the claim of character protection under copyright laws. *Beyond the Door* completed a limited theatrical run where it received universally negative reviews and a minimal box office.
43. William Van Wert, "*The Exorcist*: Ritual or Therapy?" *Jump Cut* no. 1 (May–June 1974): 4.
44. This chapter refers to the difference between the general technology underlying Dolby Stereo and the very specific technology associated with the Dolby Stereo variable-area optical format for 35 mm film, or Dolby SVA. Subsequent references to Dolby Stereo will be identified as either 70 mm Dolby Stereo or Dolby SVA for 35 mm.
45. Ioan Allen, telephone interview with the author, 23 April 2001.
46. David A. Cook, *Lost Illusions: American Cinema in the Shadow of Watergate and Vietnam: 1970–1979* (Berkeley: University of California Press, 2000), 503.
47. Although Dolby Laboratories did not advertise their Dolby Stereo technology directly to audiences, the Dolby brand "Double-D" logo was regularly accompanied by the promotional phrase "Making Films Sound Better." Hence any advertisement that featured the Dolby logo also contained the slogan emphasizing the system's effect. See the advertisement for *Star Wars*, "Dolby Makes Films Sound Better," *Variety* (8 June 1977): 31; and Paul Grainge, "Selling Spectacular Sound: Dolby and the Unheard History of Technical Trademarks," in *Lowering the Boom: Critical Studies in Film Sound*, ed. Jay Beck and Tony Grajeda (Urbana: University of Illinois Press, 2008), 251–268.

48. Michel Chion, "Une esthétique Dolby Stereo," *Cahiers du Cinéma* 329, no. 18 (November 1981): xii–xiii; my translations.
49. Ibid., xii.
50. Ibid.; emphasis in original.
51. Consider, as a point of contrast, the differences between *Star Wars* and the resonant industrial soundscapes developed by David Lynch and Alan Splet for *Eraserhead* (1977). Because Lynch and Splet valued the storytelling potential of ambient sound, they carried over their use of evocative acoustic ambience to their first foray into Dolby Stereo in *The Elephant Man* (David Lynch, 1980), not relying on any of the preconceived notions about Dolby's sound aesthetic.
52. This effect became an instant cliché for the use of the surround speakers to introduce objects from offscreen space, finding repeated use in films such as *Close Encounters of the Third Kind* (Steven Spielberg, 1977) and *Independence Day* (Roland Emmerich, 1996).
53. Gianluca Sergi, "Tales of the Silent Blast: *Star Wars* and Sound," *Journal of Popular Film and Television* 26, no. 1 (Spring 1998): 18–19.
54. Michael Pye and Lynda Myles, *The Movie Brats: How the Film Generation Took Over Hollywood* (New York: Holt, Rinehart and Winston, 1979), 135.
55. See James Corcoran and Douglas Williams, "The Recording and Re-recording of Stereophonic Sound for Wide-Screen Motion Pictures," *Journal of the SMPTE* 77, no. 12 (December 1968): 1292–1294.
56. Richard Portman, telephone interview with the author, 25 June 2000.
57. "Bass Extension of Dolby Encoded 70mm Prints" (San Francisco: Dolby Laboratories, 1981).
58. Ibid.
59. "'Star Wars' Heralds Advent of New Sound Era via Dolby Rigs," *Variety* (17 May 1978): 132.
60. "Latest Gear Opens Theatres Sound to Home Hi-Fi Level?," *Variety* (27 July 1977): 6.
61. "Dolby Prints with Stereo Sound Called as Cheap as Optical," *Variety* (18 May 1977): 95.
62. Ibid.
63. Advertisement, "Dolby Makes Films Sound Better," 31.
64. "Dolby Sound Is Receiving Plaudits for Two Films," *Boxoffice* 111 (6 June 1977): 9.
65. "In 70mm and 6-track Dolby Stereo," *International Newsletter about 70mm Film* 13, no. 62 (September 2000): 8.
66. "'Star Wars' Heralds," 132.
67. See Jay Beck, "A Quiet Revolution: Changes in American Film Sound Practices, 1967–1979" (PhD diss., University of Iowa, 2003), 187–211.
68. Ursula K. Le Guin, "Tangents: Film—*Close Encounters of the Third Kind* and *Star Wars*," *Parabola* 3, no. 1 (Winter 1978): 92–93.
69. Ibid., 92.

CHAPTER 10 THE SOUND OF STORYTELLING

1. Michel Chion, "Une esthétique Dolby Stéréo," *Cahiers du Cinéma* 329, no. 18 (November 1981): xii, my translation.

2. Charles Champlin, "In 'Days of Heaven,' Sounds of Silence," *Los Angeles Times* (17 September 1978): Calendar 1.
3. Michel Chion, *Film, A Sound Art*, trans. Claudia Gorbman (New York: Columbia University Press, 2009), 132.
4. Ibid., 346.
5. Chion, "Une esthétique Dolby Stéréo," xii.
6. Ibid., xiii.
7. Gianluca Sergi, *The Dolby Era: Film Sound in Contemporary Hollywood* (Manchester: Manchester University Press, 2004), 102.
8. Vlada Petric, "Days of Heaven," *Film Quarterly* 32, no. 2 (Winter, 1978–1979): 43.
9. Charles Schreger, "The Second Coming of Sound," *Film Comment* 14, no. 5 (September–October 1978): 36.
10. Mark Cousins, "Walter Murch," in *Projections 6: Film-makers on Film-making*, ed. John Boorman and Walter Donohue (New York: Faber and Faber, 1996), 159–160.
11. Walter Murch, "Sound Design: The Dancing Shadow," in *Projections 4: Film-makers on Film-making*, ed. John Boorman, Tom Luddy, David Thomson, and Walter Donohue (London: Faber and Faber, 1995), 245–246; my emphasis.
12. See Michael Rivlin, "Motion Picture Sound Re-recording and Mixing: Dawn of a Digital Decade," *Millimeter* 8, no. 5 (May 1980): 106, 111–115.
13. Betty Spence, "Murch: The Virtuoso of Movie Sound," *Los Angeles Times* (30 August 1981): Calendar, 29.
14. Unlike the rest of the script, which was written by John Milius and Coppola, Willard's narration was scripted by Vietnam War correspondent and author Michael Herr.
15. See Christian Metz, "Aural Objects," trans. Georgia Guirrieri, *Yale French Studies* 60 (1980): 24–32.
16. Vincent LoBrutto, *Sound-on-Film: Interviews with Creators of Film Sound* (Westport, CT: Praeger, 1994), 95.
17. Ibid., 94.
18. Ibid., 93.
19. Although the film was eventually going to be presented in Dolby Stereo, the sound team chose to use dbx noise reduction because it afforded them a far greater level of noise suppression than Dolby A-type noise reduction. Of course, it should be remembered that even though the tracks were dbx-encoded during the post-production process, the final sound track was re-encoded with Dolby noise reduction for the theatrical release.
20. Rivlin, "Motion Picture Sound Re-recording and Mixing," 115.
21. See Jay Beck with Vanessa Theme Ament, "The New Hollywood, 1981–1999," in *Sound: Dialogue, Music and Effects*, ed. Kathryn Kalinak (New Brunswick, NJ: Rutgers University Press, 2015), 123–132.
22. Ric Gentry, "Sound: Bringing the Pictures to Life," *Millimeter* 11, no. 12 (December 1983): 79.
23. Jordan Fox, "Walter Murch: Making Beaches Out of Grains of Sand," *Cinefex* no. 3 (December 1980): 52.
24. Ibid.; emphasis in original.

25. Rick Altman, "The Material Heterogeneity of Recorded Sound," *Sound Theory/Sound Practice* (New York: AFI/Routledge, 1992), 29.
26. Fox, "Walter Murch," 48.
27. Spence, "Murch," 28; my emphasis.
28. Rivlin, "Motion Picture Sound Re-recording and Mixing," 115.
29. Jay Beck, "A Quiet Revolution: Changes in American Film Sound Practices, 1967–1979" (PhD diss., University of Iowa, 2003), 147–172.
30. Walter Murch, telephone interview with the author, 23 January 2001.
31. Rivlin, "Motion Picture Sound Re-recording and Mixing," 115.
32. "We didn't have any surround information at all. It was simply left-center-right. Because the state of theaters at that time was such that we felt that using the surrounds was a little bit like giving a loaded gun to a kid. Unless there is a standard, and you can be guaranteed that they're going to be played back at the right level, either they're played back too low, and so why did you go to all of that trouble anyway because nobody hears them, or they're played back too high and they ruin the whole film-going experience. So there's a very narrow window of acceptable level for surround sound." Walter Murch, telephone interview with the author, 23 January 2001.
33. Charles Schreger, "Sounds of 'Apocalypse,'" *Los Angeles Times* (27 August 1979): Section IV, 9.
34. For more on the evolution of sound design, see Jeff Smith, "The Auteur Renaissance, 1968–1980," in Kalinak, *Sound: Dialogue, Music and Effects*, 98–106; and Beck with Ament, "The New Hollywood, 1981–1999," 107–117.
35. Randy Thom, "Designing a Movie for Sound." *iris* 27 (spring 1999): 10.
36. Mitch Tuchman and Anne Thompson, "I'm the Boss," *Film Comment* 17, no. 4 (July–August 1981): 51.
37. Larry Blake, *Film Sound Today: An Anthology from Recording Engineer/Producer* (Hollywood: Reveille Press, 1984), 11.
38. *Dolby® Background Information on Film Sound* (San Francisco: Dolby Laboratories, 1980), 1.
39. Ioan Allen, Joseph Hull, and Robert Peterson, "Stereo Sound for the Theater" (paper presented at TEA Convention '81, Scottsdale, AZ, 10–13 May 1981), 10.
40. *A Chronology of Dolby Laboratories: May 1965–May 1999* (San Francisco: Dolby Laboratories, 1999), 7.
41. LoBrutto, *Sound-on-Film*, 37.
42. Sergi, *The Dolby Era*, 64.
43. Mark Mangini, "Making Sound Career Choices," *Variety* (22 April 1998): A4.
44. See Rivlin, "Motion Picture Sound Re-recording and Mixing," 106, 111–115.
45. Chion, *Film, A Sound Art*, 131.
46. Sergi, *The Dolby Era*, 147.
47. See Jay Beck, "Citing the Sound: *The Conversation, Blow Out*, and the Mythological Ontology of the Soundtrack in '70s Film," *Journal of Popular Film and Television* 29, no. 4 (Winter 2002): 156–163.
48. See Jay Beck, "'Rewriting the Audio-Visual Contract': *Silence of the Lambs* and Dolby Stereo," *Southern Review* 33, no. 3 (2000): 273–291.

SELECTED BIBLIOGRAPHY

Allen, Ioan. "The Dolby Sound System for Recording *Star Wars*." *American Cinematographer* 58, no. 7 (July 1977): 709, 748, 761.

Altman, Rick. "24-Track Narrative? Robert Altman's *Nashville*." *Cinéma(s)* 1, no. 3 (1991): 102–125.

———. *Silent Film Sound*. New York: Columbia University Press, 2004.

———, ed. *Sound Theory/Sound Practice*. New York: AFI/Routledge, 1992.

Beck, Jay. "Citing the Sound: *The Conversation, Blow Out*, and the Mythological Ontology of the Soundtrack in '70s Film." *Journal of Popular Film and Television* 29, no. 4 (Winter 2002): 156–163.

———. "The Democratic Voice—Robert Altman's Sound Aesthetics in the 1970s." In *A Companion to Robert Altman*, edited by Adrian Danks, 184–209. Hoboken, NJ: Wiley-Blackwell, 2015.

———. "'Rewriting the Audio-Visual Contract': *Silence of the Lambs* and Dolby Stereo." *Southern Review* 33, no. 3 (2000): 273–291.

Beck, Jay, and Tony Grajeda, eds. *Lowering the Boom: Critical Studies in Film Sound*. Urbana: University of Illinois Press, 2008.

Beck, Jay, with Vanessa Theme Ament. "The New Hollywood, 1981–1999." In *Sound: Dialogue, Music and Effects*, edited by Kathryn Kalinak, 107–132. New Brunswick, NJ: Rutgers University Press, 2015.

Belton, John. "1950s Magnetic Sound: The Frozen Revolution." In *Sound Theory/Sound Practice*, edited by Rick Altman, 154–167. New York: Routledge, 1992.

Blake, Larry. *Film Sound Today: An Anthology from Recording Engineer/Producer*. Hollywood: Reveille Press, 1984.

Bobrow, Andrew C. "The Filming of *Mean Streets*." *Filmmakers Newsletter* 7, no. 3 (January 1974): 28–31.

Carroll, Noël. "The Future of Allusion: Hollywood in the Seventies (and Beyond)." *October* no. 20 (Spring 1982): 51–81.

Chion, Michel. *Audio-Vision: Sound on Screen*. Edited and translated by Claudia Gorbman. New York: Columbia University Press, 1994.

———. *Film, A Sound Art*. Translated by Claudia Gorbman. New York: Columbia University Press, 2009.

———. "Révolution douce . . . et dure stagnation." *Cahiers du Cinéma* 398 (July–August 1987): 27–32. Translated as "Quiet Revolution . . . and Rigid Stagnation" by Ben Brewster in *October* 58 (Fall 1991): 69–80.

———. *The Voice in Cinema*. Translated by Claudia Gorbman. New York: Columbia University Press, 1999.

A Chronology of Dolby Laboratories: May 1965–May 1999. San Francisco: Dolby Laboratories, 1999.

Cook, David A. *Lost Illusions: American Cinema in the Shadow of Watergate and Vietnam: 1970–1979*. Berkeley: University of California Press, 2000.

Corcoran, James, and Douglas Williams. "The Recording and Re-recording of Stereophonic Sound for Wide-Screen Motion Pictures." *Journal of the SMPTE* 77, no. 12 (December 1968): 1292–1294.

Cousins, Mark. "Walter Murch." In *Projections 6: Film-makers on Film-making*. Edited by John Boorman and Walter Donohue, 149–162. New York: Faber and Faber, 1996.

Cowie, Peter. *Coppola*. New York: Charles Scribner's Sons, 1990.

Cutts, John. "*M*A*S*H, McCloud* and *McCabe*—An Interview with Robert Altman." *Films and Filming* 18, no. 2 (November 1971): 40–44.

Denisoff, R. Serge, and William D. Romanowski. *Risky Business: Rock in Film*. New Brunswick, NJ: Transaction Publishers, 1991.

De Palma, Brian. "The Making of *The Conversation*: An Interview with Francis Ford Coppola." *Filmmakers Newsletter* 7, no. 7 (May 1974): 30–34.

"Dolby Encoded High-Fidelity Stereo Optical Sound Tracks." *American Cinematographer* 56, no. 9 (September 1975): 1032–1033, 1088–1090.

Dolby Surround Mixing Manual. San Francisco: Dolby Laboratories, 1998.

Elsaesser, Thomas. "Notes on the Unmotivated Hero—The Pathos of Failure: American Films in the 70's." *Monogram* no. 6 (October 1975): 13–19.

Farber, Stephen. "Movies from Behind the Barricades." *Film Quarterly* 24, no. 2 (Winter 1970/71): 24–33.

Fox, Jordan. "Walter Murch: Making Beaches Out of Grains of Sand." *Cinefex* no. 3 (December 1980): 42–57.

Gelmis, Joseph. *The Film Director as Superstar*. Garden City, NY: Doubleday, 1970.

Gibley, Ryan. *It Don't Worry Me: The Revolutionary American Films of the Seventies*. New York: Faber and Faber, 2004.

Grist, Leighton. *The Films of Martin Scorsese, 1963–77*. New York: St. Martin's Press, 2000.

———. *The Films of Martin Scorsese, 1978–99: Authorship and Context II*. New York: Palgrave Macmillan, 2013.

Howard, Steve. "The Making of *Alice Doesn't Live Here Anymore*: An Interview with Director Martin Scorsese." *Filmmakers Newsletter* 8, no. 5 (March 1975): 21–26.

Jacobs, Diane. "Look Back: Francis Ford Coppola's *The Rain People*." *Millimeter* 4, no. 11 (November 1976): 60–61.

Jameson, Fredric. *The Geopolitical Aesthetic*. London: BFI Publishing, 1992.

Kael, Pauline. "Bonnie and Clyde." *New Yorker* (21 October 1967): 147–171. Reprinted in *Kiss Kiss Bang Bang*, 47–63. Boston: Little, Brown, 1968.

Kass, Judith M. *Robert Altman: American Innovator*. New York: Popular Library, 1978.

Kelly, William P. "Running on Empty: Reimagining Rock and Roll." *Journal of American Culture* 4, no. 4 (Winter 1981): 152–159.

Kolker, Robert Phillip. *A Cinema of Loneliness: Penn, Kubrick, Coppola, Scorsese, Altman*. New York: Oxford University Press, 1980.

———. *A Cinema of Loneliness: Penn, Kubrick, Scorsese, Spielberg, Altman*. 2nd ed. New York: Oxford University Press, 1988.

Lev, Peter. *American Films of the '70s: Conflicting Visions*. Austin: University of Texas Press, 2000.

Levin, Sid. "The Art of the Editor: *Nashville*." *Filmmakers Newsletter* 8, no. 10 (August 1975): 29–33.

Lightman, Herb A. "The Filming of *Medium Cool.*" *American Cinematographer* 51, no. 1 (January 1970): 22–25, 58.

LoBrutto, Vincent. *Sound-on-Film: Interviews with Creators of Film Sound.* Westport, CT: Praeger, 1994.

Lowry, Ed. "The Appropriation of Signs in *Scorpio Rising.*" *Velvet Light Trap* no. 20 (Summer 1983): 41–46.

Marsh, Dave. "Schlock around the Rock." *Film Comment* 14, no. 4 (July–August 1978): 7–13.

Milne, Tom. "Not a Garbo or a Gilbert in the Bunch: Alan Pakula Talks to Tom Milne." *Sight & Sound* 41, no. 2 (Spring 1972): 88–93.

Monaco, James. *American Film Now: The People, the Power, the Money, the Movies.* New York: Plume/New American Library, 1979.

Monaco, Paul. *The Sixties: 1960–1969.* Berkeley: University of California Press, 2003.

Moussus, Fabrice. "The Art of the Boom: Nat Boxer." *Filmmakers Newsletter* 9, no. 7 (May 1976): 18–22.

Murch, Walter. "Sound Design: The Dancing Shadow." In *Projections 4: Film-makers on Film-making*, edited by John Boorman, Tom Luddy, David Thomson, and Walter Donohue, 236–251. London: Faber and Faber, 1995.

———. "The Special Sounds of *THX-1138.*" *American Cinematographer* 52, no. 10 (October 1971): 1075–1076.

Occiogrosso, Peter. "Reelin' and Rockin.'" *American Film* 9, no. 6 (1 April 1984): 44–50.

Ondaatje, Michael. *The Conversations: Walter Murch and the Art of Editing Film.* New York: Alfred A. Knopf, 2002.

Pearce, Christopher. "San Francisco's Own American Zoetrope." *American Cinematographer* 52, no. 10 (October 1971): 1002–1005, 1050–1051, 1063.

Penman, Ian. "Juke Box and Johnny Boy." *Sight & Sound* ser. 2, vol. 3, no. 4 (April 1993): 10–11.

Powell, Ann. "Scorsese and His Saint." *Millimeter* 4, no. 3 (March 1976): 30–33, 52.

Powers, James. "Martin Scorsese Seminar." *Dialogue on Film* 4, no. 7 (April 1975): 2–24.

———. "William Friedkin Seminar." *Dialogue on Film* 3, no. 4 (February/March 1974): 2–36.

Prince, David, Peter Lehman, and Ohio University students. "Film Editing: An Interview with Dede Allen." *Wide Angle* 2, no. 1 (1977): 59–69.

Pye, Michael, and Lynda Myles. *The Movie Brats: How the Film Generation Took Over Hollywood.* New York: Holt, Rinehart and Winston, 1979.

Rivlin, Michael. "Motion Picture Sound Re-recording and Mixing: Dawn of a Digital Decade." *Millimeter* 8, no. 5 (May 1980): 106, 111–115.

Schreger, Charles. "The Second Coming of Sound." *Film Comment* 14, no. 5 (September–October 1978): 34–37.

Self, Robert T. "The Sounds of *M*A*S*H.*" In *Close Viewings: An Anthology of New Film Criticism*, edited by Peter Lehman, 141–157. Gainesville: University Press of Florida, 1990.

"Sensurround." *American Cinematographer* 55, no. 10 (November 1974): 1312–1313, 1345–1353.

Sergi, Gianluca. *The Dolby Era: Film Sound in Contemporary Hollywood.* Manchester: Manchester University Press, 2004.

———. "Tales of the Silent Blast: *Star Wars* and Sound." *Journal of Popular Film and Television* 26, no. 1 (Spring 1998): 12–22.

Smith, Jeff. "The Auteur Renaissance, 1968–1980." In *Sound: Dialogue, Music and Effects*, edited by Kathryn Kalinak, 83–106. New Brunswick, NJ: Rutgers University Press, 2015.

———. *The Sounds of Commerce: Marketing Popular Film Music*. New York: Columbia University Press, 1998.

Spence, Betty. "Murch: The Virtuoso of Movie Sound." *Los Angeles Times* (30 August 1981): Calendar, 28–29.

Sturhahn, Larry. "The Art of the Sound Editor: An Interview with Walter Murch." *Filmmakers Newsletter* 8, no. 2 (December 1974): 22–25.

Telotte, J. P. "Scorsese's *The Last Waltz* and the Concert Genre." *Film Criticism* 4, no. 2 (Winter 1979): 9–20.

Thom, Randy. "Designing a Movie for Sound." *iris* 27 (Spring 1999): 9–20.

Thompson, David, ed. *Altman on Altman*. New York: Faber and Faber, 2005.

Thomsen, Christian Braad. "Monte Hellman: A Profile." *Take One* 4, no. 2 (November–December 1972): 27–31.

Uhlig, Ronald E. "The Sound of the Story." *American Cinematographer* 59, no. 8 (August 1978): 782–783, 786, 813–815.

———. "Stereophonic Photographic Soundtracks." *Journal of the SMPTE* 82, no. 4 (April 1973): 292–295.

Vogelsang, Judith. "Motifs of Image and Sound in *The Godfather*." *Journal of Popular Film* 2, no. 2 (Spring 1973): 115–135.

Webb, James E. Jr. "Multi-Channel Dialogue and Effects Recording during Film Production." *American Cinematographer* 60, no. 4 (April 1979): 368–369, 424.

Webb, Jim, with assistance from Don Ketteler. "Using the Multi-Track Format for Production Film Recording." *Recording Engineer/Producer* 11, no. 3 (1980): 110, 112, 114–117.

Wood, Robin. *Hollywood from Vietnam to Reagan*. New York: Columbia University Press, 1986.

Zuckoff, Mitchell. *Robert Altman: The Oral Biography*. New York: Vintage Books, 2010.

INDEX

ABOUT THE AUTHOR

JAY BECK is an associate professor of Cinema and Media Studies at Carleton College. He has edited two book collections—*Lowering the Boom: Critical Studies in Film Sound* with Tony Grajeda (2008) and *Contemporary Spanish Cinema and Genre* with Vicente Rodríguez Ortega (2008)—and has published extensively on cinema sound. In addition to cofounding the Sound Studies Special Interest Group of the Society for Cinema and Media Studies, he is American coeditor of the journal *Music, Sound, and the Moving Image.*

CPSIA information can be obtained
at www.ICGtesting.com
Printed in the USA
LVOW05s0046250317
528396LV00023B/290/P

9 780813 564135